Coal Mining Searches

5th edition

CONVEYANCING AND PROPERTY

A Guide to the National Conveyancing Protocol: TransAction 2001 (4th edition)
The Law Society
1 85328 731 8

Conveyancing Forms and Procedures (3rd edition)
Shelley Buckingham with Allyson Colby
1 85328 601 X

Conveyancing Handbook (10th edition, forthcoming September 2003)
Frances Silverman with Annette Goss, Peter Reekie, Michael Taylor and Bernadette Whitters
1 85328 847 0

Understanding Stamp Duty on Property
Reg Nock
1 85328 880 2

Understanding VAT on Property
David Jordan
1 85328 860 8

CONVEYANCING FORMS FROM THE LAW SOCIETY*

Seller's Property Information Form 3rd edition (PROP1)
1 85328 842 X (pack of 25)

Seller's Leasehold Information Form 2nd edition (PROP4)
1 85328 786 5 (pack of 25)

Fixtures, Fittings and Contents Form 2nd edition (PROP6)
1 85328 791 1 (pack of 25)

Completion Information and Requisitions on Title (PROP7)
8 85328 557 9 (pack of 25)

Standard Business Lease: Whole Building
8 85328 137 5 (pack of 10)

Standard Business Lease: Part of Building
8 85328 132 4 (pack of 10)

*CON 29M (2003) will not be available from the Law Society

DISTRIBUTORS OF FORM CON 29M (2003)

The form will be available from **26 May 2003** by contacting the following suppliers:

Everyform	(www.everyform.co.uk)
Laserform	(Tel. 01925 750000)
OyezStraker Office Supplies	(Tel. 01908 361166)
Peapod Solutions Limited	(Tel. 0870 380 1122)
Shaw & Sons Limited	(Tel. 01322 621100)
Stat Plus Limited	(Tel. 020 8646 5500)

COAL MINING SEARCHES

Directory and Guidance

5th edition

The Coal Authority
and
The Law Society

© The Law Society 2003

ISBN 1 85328 910 8

Published in 2003 by
The Law Society
113 Chancery Lane
London WC2A 1PL

Typeset by J&L Composition, Filey, North Yorkshire
Printed by TJ International Ltd, Padstow, Cornwall

Contents

Contents

PART I

Law Society's Guidance Notes 2003

1 INTRODUCTION

1.1 A coal mining search should be made by solicitors when acting on the occasion of any dealing with land in coal mining areas ('affected areas'), including purchase, mortgage, further advance or before any development takes place. For those solicitors not using the National Land Information Service (NLIS) or other electronic means, the search should be made using form CON 29M (2003) which is approved by the Law Society and the Coal Authority. The search should be made before the exchange of contracts or any binding obligation is entered into.

1.2 Solicitors are recommended to submit a plan of the property with every coal mining search. Plans should be marked with the full boundary of the property and not just the property building footprint or other lesser area. Solicitors should retain a copy of the search form and the plan.

1.3 These Guidance Notes should be read in conjunction with the Coal Authority's Terms and Conditions 2003 (see p.16) and User Guide 2003 (see p.3).

2 PRELIMINARY ENQUIRIES

2.1 If the property is in an affected area (see User Guide 2003, para. 2), a solicitor should make a search on form CON 29M (2003) and raise an additional preliminary enquiry of the seller. The enquiry should ask whether during the ownership of the seller, or to the seller's knowledge his predecessors in title, the property has sustained coal mining subsidence damage and if so how any claim was resolved (by making good or payment in respect of the cost of remedial, merged or redevelopment works, or otherwise).

2.2 If the mining report discloses a current stop notice or the withholding of consent to a request for preventive works affecting the property a solicitor should ask preliminary enquiries of the seller as to the present position.

3 REPRODUCTION OF FORMS

3.1 The form CON 29M (2003) and enquiries are the copyright of the Law Society which has granted to solicitors a non-exclusive licence to reproduce them. Any such form must follow precisely and in all respects the printed version (see Part V, p.79–80).

3.2 Any reproduction of the form CON 29M (2003) which does not comply with these requirements will be rejected by the Coal Authority.

4 MINING SURVEYS AND SITE INVESTIGATION

4.1 Disclosure of a disused mine shaft or adit in a mining report, the existence of recorded or possible unrecorded shallow coal workings and/or any other coal mining related hazard identified within the mining report, should be brought to the attention of the client.

If further information or advice is required in addition to that available from the Coal Authority, then solicitors should in these circumstances explain to clients that there are experienced mining surveyors and structural engineers able to advise as to what further enquiries, surveys or investigation should be made.

4.2 If a lender is involved in the transaction, solicitors should establish that the surveyor or engineer selected is acceptable to the lender.

4.3 In most cases, but not all, any shaft or adit will be owned by the Coal Authority and not the adjacent surface landowner. Clients should be advised accordingly and reminded that in these cases the permission of the Coal Authority must be sought before carrying out any works to locate, treat or in any other way interfere with former coal workings including disused coal mine shafts or adits.

5   DEALING WITH LENDERS

5.1 If domestic property which is the subject of a coal mining search is to be charged as security for a loan, a copy of the mining report should be sent to the lender as soon as received depending on the result and the lender's instructions. The solicitor should not comment substantively on the replies within the mining report but should recommend that they are referred to the lender's valuer to review.

5.2 Provided that a copy of the mining report has been so provided solicitors are not obliged to make any other reference to the replies in any mining report on title to a lender save to refer to the existence of the search and the mining report.

5.3 The Royal Institution of Chartered Surveyors, the Council of Mortgage Lenders and the Association of British Insurers have been consulted with regard to these Guidance Notes in respect of, surveys of, loans granted on security of, and insurance of, domestic properties in areas affected by coal mining, and each such organisation has prepared separate guidance to its own members.

5.4 With regard to non-domestic property a similar procedure should be adopted. Solicitors should, however, refer to the replies to the additional enquiries included in the mining reports for non-residential, commercial or development sites as these deal with legal matters (namely the withdrawal of support and the existence of working facilities orders).

5.5 When also acting for the lender solicitors should, in all cases, check whether the instructions from that lender require the solicitor to deal with the mining report in any other manner. If so, the solicitor should explain to the lender the basis upon which the solicitor is recommended by these paragraphs to proceed. It is important that solicitors should not attempt to perform the function of the client's valuer or surveyor with regard to the mining report.

6   IMPLEMENTATION OF FORM CON 29M (2003)

6.1 The new form CON 29M (2003) will be available from 26 May 2003. The Coal Authority will continue to accept the 1998 form until 1 July 2003.

6.2 From 30 April 2003, all mining reports requested from the Coal Authority will be prepared in accordance with the Law Society's Guidance 2003, the User Guide 2003 and the Terms and Conditions 2003.

PART II

User Guide 2003

1 INTRODUCTION

1.1 The past legacy and ongoing working of coal can affect surface property. Consequently, property professionals consider coal mining searches to be vital for anyone buying or developing property in any coal mining area in England and Wales.

1.2 The Coal Authority ('the Authority') holds and maintains the national coal mining database and their Mining Reports Service provides a fast, accurate and cost-effective service. Mining reports provide property-specific information about past, current and future underground and surface coal mining activities affecting any individual property or site in England and Wales.

1.3 Before making a search, users should familiarise themselves with this User Guide. For the purpose of this Guide, users should be taken to include solicitors, licensed conveyancers, surveyors, valuers, estate agents, lending organisations, insurers, surface developers and any other individual or organisation making a coal mining search with the Authority for their own or their clients use.

1.4 The Law Society recommend that a coal mining search in the form approved by the Law Society and the Authority should be made by solicitors when acting on the occasion of any dealing with land in coal mining areas ('affected areas'), including purchase, mortgage, further advance or before any development takes place. The search should be made before the exchange of contracts or any binding obligation is entered into.

1.5 This User Guide should be read in conjunction with the Authority's Terms and Conditions 2003 (see p.16).

2 AFFECTED AREAS

2.1 Though past and current coal mining activities are quite widespread, most of England and Wales are not affected areas. A coal mining search is required if the property is within an area which may be affected by previous, current or proposed working of coal. The user should not rely upon his own 'local knowledge' in determining whether or not a search should be made.

2.2 Subject to paragraph 3.2 of the User Guide, no coal mining search is required to be made in respect of property in any of the following counties of England:

Bedfordshire	Berkshire
Buckinghamshire	Cambridgeshire
Cornwall	Dorset
East Sussex	Essex
Greater London	Hampshire (including Isle of Wight)
Hertfordshire	Norfolk
Northamptonshire	Oxfordshire
Rutland	Suffolk
Surrey	West Sussex
Wiltshire	

or any of the following counties or county boroughs in Wales:

Ceredigion
Conwy
Gwynedd (excluding Isle of Anglesey (Sir Ynys Mon)).

2.3   A coal mining search is necessary in respect of properties in certain specified places in the following counties and county boroughs of England:

Bristol
Cheshire
Cumbria
Derbyshire
Devon (see paragraph 2.4 of the User Guide)
Durham
East Riding of Yorkshire
Gloucestershire (including Forest of Dean)
Greater Manchester
Hereford & Worcester
Kent
Lancashire
Leicestershire (but excluding Rutland)
Lincolnshire
Merseyside
Northumberland
Nottinghamshire
Shropshire
Somerset
Staffordshire
Tyne & Wear
Warwickshire
West Midlands
Yorkshire (North, South and West).

It is also necessary in respect of the following counties or county boroughs in Wales:

Isle of Anglesey (Sir Ynys Mon)
Blaenau Gwent
Bridgend (Pen-y-Bont Ar Ogwr)
Caerphilly (Caerffili)
Carmarthenshire (Sir Gaerfyrddin)
Cardiff (Caerdydd)
Denbighshire (Sir Ddinbych)
Flintshire (Sir y Fflint)
Merthyr Tydfil (Merthyr Tudful)
Monmouthshire (Sir Fynwy)
Neath Port Talbot (Castell-Nedd Port Talbot)
Newport (Casnewydd)
Pembrokeshire (Sir Benfro)
Powys
Rhondda Cynon Taff (Rhondda Cynon Taf)
Swansea (Abertawe)
Torfaen (Tor-Faen)
Vale of Glamorgan (Bro Morgannwg)
Wrexham (Wrecsam)

The procedures for finding out which places within the above county/county boroughs require a search to be made are outlined at paragraph 3.1 of the User Guide.

4

2.4 It has recently been discovered that in an area of countryside between Bideford and Chittlehampton in Devon there may have been coal workings, perhaps associated with clay mining, and some 20 disused shafts or adits. In these circumstances, Abbotsham, Bideford, Chittlehampton, Clapworthy, East the Water, Hiscott, Umberleigh and Woodtown are now places in Devon where a coal mining search is required.

2.5 Not all property located within an affected area is within the zone of likely physical influence on the surface of underground coal working. Property which is within such a zone will not necessarily sustain subsidence damage but some support from the surface where the property is situated may have been or may be withdrawn in the future. Calculations relating to the likely zone of influence on the surface from mining activities will be based on the principle of 0.7 times the depth of the working allowing for seam inclination.

2.6 Licensed mine operators are required to give property owners advance notice of any proposals for underground coal mining operations which might result in subsidence affecting the property and of any decision not to proceed with the operations or anything which gives them reason to believe there is no longer any risk of the property being affected by subsidence and of the discontinuance of any operations which have been carried on.

3 PROCEDURES FOR FINDING OUT WHETHER A COAL MINING SEARCH IS NECESSARY

3.1 Users can establish whether a coal mining search is required by:

1. Using the Directory of Places in Part IV – an updated version of which is maintained on the Authority's corporate website at www.coal.gov.uk.

2. Using the Authority's Online Directory (www.coalminingreports.co.uk) which provides immediate online confirmation on entering a property postcode as to whether a coal mining search is necessary or not. By its nature, this electronic referral system is more precise than using a printed listing of place names.

3.2 When the Authority state, by whatever means, that no coal mining search is required, this does not and should not be deemed to confirm that no coal mining strata is present, nor that some part of any coal resources present will (subject to obtaining planning permission and a licence from the Authority) not be worked at some future date.

4 HOW TO USE THE PUBLISHED DIRECTORY LISTINGS

4.1 The user should inspect the Directory of Places (see p.22) to ascertain whether or not the name of the place in which the property is situated is listed in bold type. If the place appears in bold type then a coal mining search should be made. If a place is listed but not in bold type then no search is required. Where a place is not listed, users should contact the Authority on 0845 762 6848 or by email at miningreports@coal.gov.uk.

5 HOW TO USE THE AUTHORITY'S ONLINE DIRECTORY

5.1 To use the Authority's Online Directory:

1. Log onto the Authority's website at www.coalminingreports.co.uk (a link is also available from the corporate website at www.coal.gov.uk).

2. Enter the site by either the regular user (see paragraph 10.2 of the User Guide) or non-regular user route as required.

3. Enter the house number or name and postcode in the boxes provided.

4. Confirm the corrected Royal Mail Address File (PAF) search address when prompted.

5. You will be immediately advised that, from the information available to the Authority, either:

    (a) a coal mining search is not required for the property (with the option to purchase a certificate to confirm this information, if required); or

    (b) that a coal mining search is recommended (and given the option of mining report types and services available for purchase) and the facility to make searches online.

6. If users do not wish to make a search online they can exit the service at this stage, or search against other property postcodes.

6 SEARCH ENQUIRIES AND TYPES OF MINING REPORT

6.1 Two types of search can be made (see paragraph 14 of the User Guide for fees):

1. **Residential Property Search.** This is available for single unit residential property, either existing or currently being built, i.e. having already been the subject in full or part of a previous development site search. By way of illustration, this includes any single unit domestic property (e.g. a house, flat or bungalow including any associated garage or car-parking space), a single plot on a multi-plot development site, a farmhouse or similar converted property (*but excluding any associated outbuildings and/or land*). The mining report answers enquiries 1 to 9 of form CON 29M (2003), relating to:

    • past, present and future underground coal mining;
    • shafts and adits;
    • surface geology;
    • past, present and future opencast coal mining; and
    • subsidence (damage notice/claim/method of discharge of any remedial obligations/stop notice/request for preventive works).

    Replies to these searches will assume that no development will take place.

2. **Non-Residential Property or Development Site Search.** This is available for non-domestic properties, or properties where development is intended. By way of illustration, non-residential, commercial or development sites include multiple residential property requests (e.g. a pair of semi-detached houses or a row of terraced houses), public houses, shops, businesses, commercial property, industrial estates, rural estates, pipelines, roads and similar linear structures up to 500 metres in extent, *any* sized development site from a single plot up to a maximum of 25 hectares in extent, farmhouses and associated outbuildings and land (see also pararaph 14.2 of the User Guide). The mining report answers enquiries 1 to 12 of form CON 29M (2003), including the Residential Property Search enquiries 1–9 and the additional enquiries relating to:

    • withdrawal of support;
    • working facilities orders; and
    • payments to owners of former copyhold land.

Replies to these searches will assume that development may take place.

6.2 With regard to withdrawal of support, users should be aware that the common law principles giving surface landowners a prima facie right of support may be over-ridden by statutory provisions. Such statutory provisions are contained in coal mining legislation such as:

- Coal Industry Act 1994, s.38;
- Coal Industry Act 1975, s.2; and
- Coal Act 1938, Schedule 2, paras 5 and 6.

The mining report will give details of any notice(s) given under the relevant legislation. The mining report will not give details of any rights to withdraw support contained in title or severance deeds. Whether any support has or may be withdrawn can be answered only by reference to the records presently available to the Authority and, where appropriate, will be effectively answered under the past, present and future headings of the mining report reply.

7 OVERVIEW OF SEARCH METHODS

7.1 Coal mining searches can be made electronically either through a National Land Information Service (NLIS) channel provider or using the Authority's own Online Service. Searches can also be made by post, fax (for expedition) or telephone. Details of each method are outlined in paragraphs 8 to 13 of the User Guide.

7.2 The majority of searches made using NLIS, Coal Authority Online Service, telephone and fax are returned within 24 to 48 hours. Most postal searches are dealt with within five working days of receipt. Whilst these turnaround times are typical, they cannot be guaranteed. Searches should therefore be made as early as possible in the conveyancing process to avoid the consequence of any delay. This will ensure timely return of mining reports and avoid potential difficulties in meeting the user's own time schedules.

8 ORDERING ELECTRONICALLY

8.1 This is the quickest, easiest and most convenient way for users to make searches and receive mining reports. Two electronic methods are available:

(a) using the National Land Information Service (NLIS) through which other searches can be made; and

(b) using the Authority's own Online Service.

9 NATIONAL LAND INFORMATION SERVICE (NLIS)

9.1 This is the most efficient way for users to make searches and receive mining reports. The government has licensed three private sector organisations to operate the NLIS 'channel' services via the Internet. Contact details for the three NLIS channel providers are given at para. 9.2 below. Each channel provides an online link to the Authority's Online Service. Users should contact their preferred channel provider direct for advice on how to make searches.

9.2   The NLIS channel providers are:

MacDonald Dettwiler (Channel) Ltd
Ground Floor, Clare House
Langley Business Centre
Station Road
Langley
Berkshire SL3 8DS
Tel: 0800 085 4951 (Customer Services)
Fax: 01753 214 599
Email: support@transaction-online.co.uk
Website: www.transaction-online.co.uk

NLIS Searchflow
Crown House
Home Gardens
Dartford
Kent DA1 1DZ
Tel: 0870 755 9940 (Customer Sales)
Fax: 0870 990 9949
Email: searchflow@searchflow.co.uk *or* sales@searchflow.co.uk
Website: www.searchflow.co.uk

TM Property Service Ltd
Delta 200
Delta Business Park
Swindon
Wiltshire SN5 7XP
Tel: 0870 740 7833 (Helpdesk)
Fax: 0870 741 0426
Email: info@tmproperty.co.uk
Website: www.tmproperty.co.uk

9.3   Users should ensure that the full boundary of the property is provided to the Authority and not just the property building footprint or other lesser area.

10   COAL AUTHORITY'S ONLINE SERVICE

10.1   In order to use the Online Service users will require the following:

- the postcode and house number or name of the property;
- an email address;
- contact details including name, address and telephone number;
- a credit or debit card (unless a monthly invoiced account is used);
- Netscape browser version 4.0 or above/Microsoft Internet Explorer browser version 5.0 or above; and
- a screen resolution of 800 × 600 pixels or above.

10.2   Regular users can apply for a user ID and password to access the system. Application forms for users to become registered users of the Authority's Online Service are available from the Authority. This enables users to enter the system using a user ID and password, and for mining reports to be dispatched using pre-recorded address details. Registered users are allocated at the discretion of the Authority. If users wish to find out about eligibility to become a registered user, they should contact the Authority on 0845 762 6848 or at miningreports@coal.gov.uk.

10.3 Mining reports and certificates will be dispatched to users via email as a PDF attachment. Users must have the Adobe Acrobat Reader to view these attachments (if you do not have this it can be down-loaded from www.adobe.co.uk for free).

10.4 The majority of mining reports are returned within 24 to 48 hours, the remainder within 72 hours.

10.5 Payment for mining reports is made via the WorldPay secure payment system or by monthly invoice. A monthly invoiced account arrangement is available but its use is subject to application, acceptance and a formal credit agreement with the Authority.

10.6 Where the WorldPay secure payment system is used, the Authority's Online Service will automatically link users to the WorldPay system when payment is required. For security reasons, no credit or debit card details are seen or stored by the Authority, and all payment transactions take place directly within the WorldPay system. For more information on the WorldPay system see the Authority's or WorldPay website (www.world pay.com).

10.7 The Authority's Online Service is based on postcode searching and is therefore not suitable for parcels of land, new or unbuilt properties. However, mining reports for such properties can be returned by post, fax or email. Contact the Authority on 0845 762 6848 for further details.

10.8 Once payment is confirmed by WorldPay, or account holders receive confirmation that mining reports have been added to monthly invoices, reports or certificates will be automatically produced by the Authority. Users will therefore not be able to cancel orders after receiving confirmation that mining reports or certificates have been ordered.

11 SEARCHES BY TELEPHONE

11.1 Account and credit/debit card users can also make searches by telephone on 0845 762 6848. Users then have the option of having the mining reports returned by post, fax or email. The procedure is as follows:

11.2 Before calling, users need to have the following information to hand.

Account holders will need:

- customer account number; and
- the house number and postcode of the subject property (or full site description for non-addressable property).

Non-account holders will need:

- credit/debit card details;
- contact telephone number;
- the house number and postcode of the subject property (or full site description for non-addressable property); and
- contact details of the preferred method of receiving the mining report(s) (i.e. postal or email address or fax number).

11.3 Users should call the Authority on 0845 762 6848 and inform the Helpline search assistant that they wish to use the Authority call centre service to order a mining report(s).

11.4 The search assistant should be advised as to whether the user is an account holder or not. If so, a customer account number must be provided.

11.5 The search assistant will ask for the house number and postcode of each subject property and confirm the Royal Mail's Address File (PAF) corrected address. The facility is also available to log a non-PAF address (e.g. for a development plot or site) but users should forward a plan separately by post, fax or email (at the users preference) delineating the boundaries of the site (see paragraph 11.8 of the User Guide).

11.6 The search assistant will confirm account holders' preferred method of receiving the mining report(s). Non-account holders will be asked for their preferred return route and asked for details, i.e. postal or email address or fax number.

11.7 The search assistant will digitise the property boundary, seeking any clarification as necessary.

11.8 Where the extent of the property boundary needs further clarification, users will be asked to provide a location plan by post, fax or email (at the users preference) and given a reference number to quote.

11.9 Multiple requests can be accommodated.

11.10 For non-account holders, the search assistant will advise of the total fee for the mining report(s) ordered and request and input the user's credit/debit card details.

11.11 On successful credit/debit card authorisation, boundary digitisation will be confirmed and the mining report(s) will be passed for processing.

11.12 If credit/debit card authorisation is unsuccessful, the process cannot proceed and the user will be asked to request the mining report(s) by another method, e.g. by post (see paragraph 12 of the User Guide).

12   SEARCHES BY POST

12.1 Searches can be made by post using the Law Society form CON 29M (2003). Copies of CON 29M (2003) can be purchased from Everyform, Laserform, OyezStraker Office Supplies, Peapod Solutions Limited, Shaw & Sons Limited, and Stat Plus Limited.

12.2 The search form must be completed by insertion of the full postal address including the postcode of the property. As full a description as possible should be given where the property address is not of the standard single number and postcode type. The search form should include the name and either the document exchange (DX) number of the user, or postal address, and an email address if electronic return of the mining report is preferred. The request should also include the user's file reference and telephone number. Mining reports returned by email will be in secure PDF format. Users should indicate the mining report type required, i.e. whether it is for a residential property or for a non-residential or development site. It is inappropriate to request a residential property search for non-residential property or development sites and any such requests will be returned for resubmission and the correct fee. The search form also contains provision for the search to be expedited by fax on payment of an additional fee (see paragraph 13 of the User Guide).

12.3 Form CON 29M (2003) should be sent with a plan of the property and the appropriate fee to:

The Coal Authority
Mining Reports
200 Lichfield Lane
Mansfield
Nottinghamshire NG18 4RG
DX 716176 Mansfield 5

12.4 When searching by post, users are recommended to submit a plan with every search application as it is likely to be replied to sooner than a search application without a plan. If a plan is not submitted there may be difficulties and delays in identifying the property or its extent, in which circumstances, the Authority may request that a plan be supplied with consequent delay in replying to the search application. When submitting requests for development site plots, the accompanying plan should show the plot boundary in relation to established surface features. Plans will be retained by the Authority.

12.5 No covering letter is required to be sent nor is it necessary to sign the search form. A copy of the search and plan should be retained and affixed to the replies when received. If photocopies of the form are used it is important that these are of good quality and both sides of the search form should be reproduced for retention and annexing to the reply when received.

12.6 A separate search is required in respect of each individual property.

12.7 Poor quality photocopies of the search form may not be accepted and will be returned.

12.8 Users are asked:

- not to staple anything to the form – cheques should be attached with a paper-clip or similar;
- not to send a covering letter; and
- to write clearly – the form should be typed or written in block letters.

Users are asked to *ensure* they:

- include a postcode where available;
- complete all the entries on the form; and
- enclose and sign an accompanying cheque and ensure it is for the correct fee.

13 EXPEDITED SEARCHES

13.1 Notwithstanding the introduction of the NLIS and the Coal Authority Online Service (which return the majority of mining reports within 24 to 48 hours), the Authority's Expedited Search service is still available, on payment of an additional fee.

13.2 In using this service a user may make a search by fax and receive a mining report by fax with 48 hours. An additional fee is payable for expedition and this can be established by contacting the Authority's helpline on 0845 762 6848 or visiting their website at www.coalminingreports.co.uk.

13.3 In order to make an expedited search the form should be completed as normal but the request for expedition of the reply (which constitutes an undertaking to pay the expedition fee within seven days) should also be completed. The form and plan should then be faxed to the Authority on 01623 638338. The user should, on the same day, send the requisition form with the fee (including the additional fee for expedition) to the Authority by

11

post or DX. In normal circumstances the replies will be returned by fax the same day or the next working day.

13.4   The Authority will then, on receipt of the confirmatory requisition and fee, send the confirmatory replies with the VAT receipt for the fee.

14   FEES

14.1   The current scale of charges is available on the Authority's website at www.coalminingreports.co.uk. Advice is also available on contacting the Authority's helpline on 0845 762 6848.

14.2   The actual fee payable for a coal mining report will depend on the method of ordering and whether it is for a residential property or for a non-residential or development site up to a maximum of 25 hectares in extent. Additional fees are payable for sites in excess of 25 hectares in extent and for linear property, like pipelines and roads etc. that exceed 500 metres in length. Certificates to confirm that no search is required (see paragraph 5.1 of the User Guide) attract a lesser fee.

14.3   The fees charged in respect of searches – both for residential and for non-residential, and development sites – and for expedition, are reviewed from time to time. The Law Society will be consulted before any change is proposed. Notification by mailshot letter or the publication on the Authority's website of a revised scale of charges constitutes notice of any fee change. Changed fees become payable from the date advised on any notification letter or website scale of charges.

14.4   For postal searches, the fee is payable when the search is made and should accompany the form. The mining report will contain a VAT receipt for the fee. The VAT element of the fee should be treated by the user as an input for VAT purposes and VAT must be charged to the client. It should not be necessary to retain the receipted reply (nor a copy) for VAT purposes.

14.5   If, by mistake, a user makes a coal mining search in respect of property outside the affected areas, the Authority will reply but the fee will not be refunded.

14.6   The Authority will return any requests received that enclose an incorrect fee with a request for the correct fee.

14.7   The Authority cannot guarantee to cancel a search once a request has been made. No refund or transfer of any fee (or part thereof) will be made once a search has been requested.

15   PAYING FOR COAL MINING REPORTS

15.1   Depending on the method used to search and to receive mining reports, various methods of payment are available. Payment can be made:

- by cheque in advance (for all postal searches) made payable to the Coal Authority;
- by credit/debit card (telephone and online customers only); and
- by customer account (telephone and online customers only).

16 CUSTOMER ACCOUNT FACILITIES

16.1 Customer account facilities are available (subject to conditions) for users of the Authority's Online or telephone services. To apply for a monthly invoiced account users will need to complete an application form which is available from the Authority on 0845 762 6848 or by email at miningreports@coal.gov.uk. Accounts will only be offered to those customers who satisfy the selection criteria and are subject to a formal credit agreement with the Authority. No account facility will be provided to users who are currently on the stop list of either the Authority's MRSDS system or the expedited service. Account facilities will only be provided to those users who request, on average, ten or more coal mining reports each month.

16.2 Unless payment is made in advance, the Authority will issue an invoice to the user account holder at the end of each calendar month, or as soon as is practical thereafter, for payment for all coal mining reports dispatched to the user account holder during that month. Payment must then be made by the user account holder to:

The Coal Authority
Mining Reports
200 Lichfield Lane
Mansfield
Nottinghamshire NG18 4RG
DX 716176 Mansfield 5

by the end of the month following the month in which the mining report was dispatched ('the due date').

Any sums outstanding after the due date will bear interest, at the rate of five per cent per annum above base rate, from the date of the invoice to the date of actual payment. In the event that payment is not made within 30 days from the date of the invoice, the Authority will have the absolute right to cancel the credit account facilities.

16.3 The Authority reserve the right to withdraw or suspend customer account facilities where the volumes of mining reports requested falls below, on average, ten per month or for persistent late payment of invoices.

16.4 All NLIS users will have customer account facilities with their preferred NLIS channel provider (as opposed to the Authority) via whom the Authority will be reimbursed in respect of mining reports delivered through NLIS.

17 PLANS

17.1 Users should ensure that the full boundary of the property is provided to the Authority and not just the property building footprint or other lesser area.

17.2 The Authority provides with each mining report a plan of the boundaries of the property in respect of which the mining report has been prepared. It is necessary for users to ensure that such boundaries correspond with those of the property. If the boundary of the property shown on the Authority plan does not so correspond, any discrepancy should be resolved with the Authority by users. If the discrepancy is not referred to the Authority by users within 28 days of the date of the mining report in question, users will have to make a fresh search with payment of the appropriate fee. The property will be located with reference to Ordnance Survey (OS) digital publications in accordance with the plan updates agreement between OS and the Authority. The Authority cannot and does not warrant that the OS information is complete or accurate and accepts no liability for the plotted position of property as shown on published OS maps.

17.3 Ordnance Survey is undertaking a positional accuracy improvement programme of its mapping data. The Authority has no control over the timing of issue of positionally improved mapping data by OS or over users of that data. In some instances the relative position between surface features and coal mining features may alter as a consequence of this programme. The Authority is involved in ensuring that the integrity of its database is maintained by replotting certain mining information to ensure that the relationship between that mining information and the improved OS surface positions is accurate. However, mining reports may still be produced against pre-improved OS mapping whilst the Authority updates its database in line with the OS changes.

18   SHAFTS AND ADITS

18.1 The reply as to shafts and adits and other entries to underground coal mine workings (within 20 metres of the boundary of the property) will be prepared only from the records in the possession of the Authority. These records may not be complete. The reporting distance of 20 metres is recommended and agreed with the Law Society, Royal Institution of Chartered Surveyors, Council of Mortgage Lenders and the Association of British Insurers. The approximate location of any such shaft or adit will be identified on a plan with the mining report at no extra cost. For reasons of clarity, mine entry symbols may not be drawn to the same scale as the plan. Distances are measured from the estimated centre of the shaft or centre point of an adit entrance.

18.2 References to a coal mine shaft within 20 metres of a property or its boundary does not necessarily mean that the property has or will have any instability problems. The number of cases where shafts affect the structure of a property are very low.

18.3 Users are reminded that with effect from 31 October 1994 British Coal's interests in unworked coal and coal mines became vested in the Authority. In most cases, but not all, any shaft or adit will be owned by the Authority and not the adjacent surface landowner. In these cases the permission of the Coal Authority must be sought before carrying out any works to locate, treat or in any other way interfere with disused coal mine shafts or adits.

19   SUBSIDENCE DAMAGE CLAIMS

19.1 If further subsidence damage claims information is required in addition to that provided in a mining report, the Authority will need to manually search their records. Such further enquiries should be made under separate cover to:

The Coal Authority
Claims History Service
200 Lichfield Lane
Mansfield
Nottinghamshire NG18 4RG
DX 716176 Mansfield 5

For advice on the current fee for this service contact the Authority's helpline on 0845 762 6848 or check the Authority's website at www.coalminingreports.co.uk. Alternatively, it may be that such information can be obtained by other means such as preliminary enquiries of the present owner.

20 TIME VALIDITY

20.1 There is no time protection afforded by mining reports. Whether a user can rely on a past mining report (of his own or another) depends upon all the circumstances of the case including how recently it was made, the content of the mining report, the nature of the property and the user's intentions in relation to it. Licensed operators' plans for mining may change as may the other relevant information available to the Authority. If there is any doubt as to whether a previous mining report remains valid, it is suggested that a new search should be made. Without prejudice to the generality of the foregoing, any mining report must not be relied upon in any event in excess of 90 days from its date of issue and must thereafter be verified as still being up-to-date, either by making a fresh search or by using any available update service provided by the Authority.

PART III

Terms and Conditions 2003

1 GENERAL

1.1 Coal Mining Reports ('mining reports') will be provided subject to these terms and conditions and in accordance with the duties of the Coal Authority ('the Authority') under the Coal Industry Act 1994. Mining reports will be based on, and limited to, records in the possession of the Authority. The records available to the Authority are constantly updated and added to the relevant computer database. The Authority will make use of the most up-to-date records available at the time of giving mining reports. However, no warranty is given or representation made that such records will not become obsolete or incorrect over any period of time. Mining reports will be given in the belief that they are in accordance with the information available to the Authority at the time of giving the mining report but on the distinct understanding that the Authority is not legally responsible for them except for negligence.

1.2 It is the responsibility of the person by or for whom a mining report is requested to specify the property for which a mining report is required and to make clear the full extent of the property boundary (i.e. buildings and associated land).

1.3 These Terms and Conditions of the Authority incorporate the User Guide 2003 and have been approved by the Law Society and apply to all searches made including those using form CON 29M (2003).

2 RECORDS

2.1 The records in the possession of the Authority are derived from a number of sources and are of various ages, scales, condition, etc. As a consequence information sourced from such records is of variable reliability. Additionally, in any particular area there may be information held by others as to historical coal (and other) mining which is not also in the possession of the Authority. In these circumstances the Authority are unable to give any warranty and make no representation that the information comprised in the records in its possession is complete, accurate, exhaustive or reliable.

2.2 Some records are derived from licensed operators whose plans for current and future coal mining may change at any time. They are required by s.58 of the Coal Industry Act 1994 to exercise all due diligence to secure the provision of full and accurate information to the Authority in accordance with the provisions of the licences which they hold.

2.3 In certain instances (usually relating to older records in the custody of the Authority) it has been necessary for the Authority to make assumptions as to the most probable ('best plot') positions of mine entries, the depth, date and extent of coal workings, the number and inclination of seams and the nature of the mineral worked, e.g. in certain areas records attribute more than one name to a single seam.

3   TITLE DEEDS, MINING LEASES, SEVERANCE INSTRUMENTS AND AGREEMENTS

3.1   Mining reports will not refer to or deal with rights under title deeds nor the existence (or the relevance to any claim affecting the property) of coal mining leases, instruments of severance or agreements with licensed operators, the Authority, British Coal Corporation, National Coal Board or the Coal Commission which may provide the basis of an alternative entitlement to compensation or repair or otherwise affect the position of enquirers.

4   OTHER MINERALS

4.1   Mining reports relate only to coal and minerals worked in association with coal. The presence of workings of other minerals will not necessarily be disclosed. The enquirer may need to make separate enquiries regarding other minerals to the appropriate sources of information in certain areas where these other minerals are known to exist or have been worked.

5   LIABILITY

5.1   Any liability of the Authority for negligence in giving mining reports shall be for the benefit of not only enquirers but also a person (being a purchaser for the purpose of s.10(3) of the Local Land Charges Act 1975) who or whose agent had knowledge before the relevant time (as defined in that section) of the contents of the mining report. Such extension of liability to another (who did not purchase the mining report from the Authority) is limited to a purchaser, lessee or mortgagee of the property and not others (e.g. other recipients of reports on title, etc.).

5.2   There is no time protection afforded to mining reports. Whether a person can rely on a past mining report of a particular property (whether purchased by that same person or another) depends upon all the circumstances of the case, including: how recently it was purchased, its contents, the nature of the property and the person's intentions in relation to it (including development). Licensed operators plans for mining may change as may other relevant information available to the Authority. If there is any doubt as to whether a previous mining report remains valid a new mining report should be purchased. Without prejudice to the generality of the foregoing, any mining report must not be relied upon in any event in excess of 90 days from its date of issue and must thereafter be verified as still being up-to-date, either by ordering a fresh mining report or by using any available update service provided by the Authority.

5.3   The Authority shall be deemed not to know the purpose for which mining reports are required even if such purpose is made known to them. The Authority makes no warranties or representations as to either the suitability of land/property for any particular use or purpose or its value, and shall not in any circumstances be liable for any loss or damage at all arising from reliance on mining reports in relation to these matters.

5.4   Property owners may have the benefit of remedies under the Coal Mining Subsidence Act 1991, which contains provisions relating to the making good, to the reasonable satisfaction of the claimant, of physical damage from coal mine workings, including disused coal mine entries. A DTI leaflet setting out the rights and obligations of the person responsible for subsidence damage under the 1991 Act and otherwise can be obtained by telephone on 0845 762 6848. All calls will be charged at the local rate.

6   SHAFTS AND ADITS (MINE ENTRIES)

6.1   Information in mining reports relating to mine entries (within 20 metres of the boundary of the property) will be prepared only from the records in the possession of the Authority. These records may not be complete. The approximate location of any such mine entries will be identified on a plan within the body of mining reports. To aid clarity the mine entry symbols are not necessarily shown to the same scale as the plan. Distances are measured from the estimated centre point of each recorded mine entry.

6.2   With effect from 31 October 1994 British Coal's interests in unworked coal and coal mines became vested in the Authority. In most cases (but not all) any mine entries will be owned by the Authority and not the adjacent landowner.

7   SURFACE FAULTS

7.1   No comment is made in mining reports about the existence of conjectured surface fault positions shown on geological maps (published by the British Geological Survey) which are not known to the Authority to have affected the stability of the property as a result of coal mining activities.

8   MISCELLANEOUS

8.1   During the production of mining reports by the Authority, the address (but not the boundary) of the subject property as provided by the enquirer may be corrected to match the Royal Mail's Address File (PAF).

8.2   In reporting whether or not a property lies within a former opencast site from which coal had been extracted by opencast methods, depending on the age and source of information available to the Authority, the opencast site boundary may be limited to the specific area where coal was believed to have been extracted and not to the overall site boundary or excavation area.

8.3   Where the Authority is aware that property the subject of the mining report has been affected by mine gas, information will be included in the mining report, together with details, where available, of any remedial works carried out by the Authority.

9   PLANS

9.1   The Authority provides with each mining report a plan of the boundaries of the property in respect of which the mining report has been prepared. It is necessary for enquirers to ensure that such boundaries correspond with those of the property. If the boundary of the property shown on the Authority plan does not so correspond, any discrepancy should be resolved with the Authority by enquirers. If the discrepancy is not referred to the Authority by enquirers within 28 days of the date of the mining report in question, enquirers will have to make a fresh search with payment of the appropriate fee. The property will be located with reference to Ordnance Survey (OS) digital publications in accordance with the plan updates agreement between OS and the Authority. The Authority cannot and does not warrant that the OS information is complete or accurate and accepts no liability for the plotted position of property as shown on published OS maps.

9.2   Information supplied in plan form within the body of mining reports should not be enlarged to any greater scale than that at which it is originally supplied or accuracy will be affected.

10   MINING SURVEYS AND SITE INVESTIGATIONS

10.1   A mining report is not a substitute for site investigation or a mining survey. Enquirers will have to assess whether a site investigation or mining survey is required having regard (amongst other relevant factors) to the content of the mining report and whether the property is to be developed and, if so, the nature and extent of the development. There are experienced mining surveyors and structural engineers in all coal mining areas able to advise as to what further enquiries, mining surveys or site investigations should be made.

11   COAL MINING REPORT INFORMATION

11.1   In respect of each residential property search, mining reports will provide summary information according to the records in the possession of the Authority relating to those matters, referred to below on the basis of and subject to the terms and conditions referred to in paragraphs 1 to 10 inclusive above. For each search the mining report will include the following information:

Past underground coal mining

- Whether the property is within the zone of likely physical influence on the surface of past underground working based on the principle of 0.7 times the depth of the working, allowing for seam inclination (the Authority will indicate the number of seams involved, minimum and maximum depth and the approximate last date of working).

- An indication of whether ground movement should now have ceased when the circumstances are considered appropriate.

- An indication of the likely existence of unrecorded coal workings.

Present underground coal mining

- Whether the property is within the zone of likely physical influence on the surface of present underground coal workings based on the principle of 0.7 times the depth of the working, allowing for seam inclination (the seams involved are indicated).

Future underground coal mining

- Whether the property lies within the geographical area for which a licence to extract coal by underground methods is awaiting determination by the Authority or is extant (the date an extant licence was granted is also given) together with advice as to whether the licence is conditional.

- Whether the property is within the zone of likely physical influence on the surface based on the principle of 0.7 times the depth of the currently planned future underground working, allowing for seam inclination (an indication will be given of the seams involved and approximate date of working).

- Information as to whether further workable coal is known or thought to exist.

- Whether any notice of proposals relating to underground coal mining operations have been given under s.46 of the Coal Mining Subsidence Act 1991 (as amended

19

by the Coal Industry Act 1994), and if so details are given of the date of the last notice.

Shafts and adits

- Details of any shafts or adits within the property and/or within 20 metres of the boundary of the property.

- Brief treatment details are given, where known.

- A plan showing the approximate location of any shafts and/or adits is provided.

Surface geology

- Whether the Authority have record of a fault or other line of weakness that is known to the Authority to affect the stability of the property.

Past opencast coal mining

- Whether the property lies within an opencast site boundary from which coal has been extracted in the past by opencast methods.

Present opencast coal mining

- Whether the property is within 200 metres of the boundary of an opencast site from which coal is being extracted by opencast methods.

Future opencast coal mining

- Whether the property is within 800 metres of an area for which an application for a licence to extract coal by opencast methods is awaiting determination by the Authority.

- Whether the property is within 800 metres of an area for which a conditional licence to extract coal by opencast methods has been granted by the Authority.

Subsidence

- The date of any damage notice or claim made or pursued for alleged coal mining subsidence damage since 1 January 1984.

- Whether the claim was accepted, rejected or whether liability is still being determined.

- Where a claim has been discharged, whether this was by making good or by payment of compensation or a combination of both. No details of the works to be made good or of the amount paid will be given (additional information may be available, subsequent to receipt of a mining report, on written request and payment of an additional archive research fee).

- Whether there is any current 'stop notice' concerning the deferment of remedial works or repairs affecting the property, and if so the date of the notice.

- Whether there has been any request made for preventive works under s.33 of the Coal Mining Subsidence Act 1991; if yes, has any person withheld consent or failed to comply with any request to execute preventive works?

11.2 Reports for non-residential or development sites will also include the following additional information:

Withdrawal of support

- Whether the site lies within an area in respect of which a notice of entitlement to withdraw support has been published (the date of any notice is provided).

- Whether the site lies within an area in respect of which a revocation notice has been given under s.41 of the Coal Industry Act 1994 (the date of any notice is provided).

Working facilities orders

- Whether the site lies within an area affected by an order in respect of the working of coal under the Mines (Working Facilities and Support) Acts 1923 and 1966 or any statutory modification or amendment thereof (the date of any such notice is provided).

Payments to owners of former copyhold land

- Whether any relevant notice which may affect the property has been given and, if so, details of any notice of retained interests in coal and coal mines, acceptance or rejection notices and whether any compensation has been paid to a claimant.

PART IV
Directory of Places

A

AB KETTLEBY (Leicestershire)
Ab Lench (Hereford & Worcester)
Abbas Combe (Somerset)
ABBERLEY (Hereford & Worcester)
ABBERLEY COMMON (Hereford & Worcester)
Abberton (Hereford & Worcester)
Abberwick (Northumberland)
Abbey Dore (Hereford & Worcester)
Abbey Green (Staffordshire)
Abbey Hill (Somerset)
Abbey Town (Cumbria)
ABBEY VILLAGE (Lancashire)
Abbeycwmhir (Powys)
ABBEYDALE (South Yorkshire)
Abbeystead (Lancashire)
ABBOT'S CHAIR (Derbyshire)
Abbot's Salford (Warwickshire)
Abbots Bromley (Staffordshire)
Abbots Leigh (Somerset)
Abbots Morton (Hereford & Worcester)
ABBOTSHAM (Devon)
Abcott (Shropshire)
ABDON (Shropshire)
Abenhall (Gloucestershire)
Aber Clydach (Powys)
Aber-arad (Carmarthenshire)
Aber-giar (Carmarthenshire)
ABER-NANT (Rhondda Cynon Taff)
ABERAMAN (Rhondda Cynon Taff)
Aberangell (Powys)
ABERAVON (Neath Port Talbot)
ABERBARGOED (Caerphilly)
ABERBEEG (Blaenau Gwent)
ABERCANAID (Merthyr Tydfil)
ABERCARN (Caerphilly)
Abercastle (Pembrokeshire)
Abercegir (Powys)
ABERCRAF (Powys)
ABERCREGAN (Neath Port Talbot)
ABERCWMBOI (Rhondda Cynon Taff)
Abercych (Pembrokeshire)
ABERCYNON (Rhondda Cynon Taff)
ABERDARE (Rhondda Cynon Taff)
ABERDULAIS (Neath Port Talbot)
Aberedw (Powys)
Abereiddy (Pembrokeshire)
ABERFAN (Merthyr Tydfil)
Aberffraw (Anglesey)
ABERFORD (West Yorkshire)
ABERGARW (Bridgend)
ABERGARWED (Neath Port Talbot)
Abergavenny (Monmouthshire)
Abergorlech (Carmarthenshire)
Abergwesyn (Powys)
Abergwili (Carmarthenshire)
Abergwydol (Powys)
ABERGWYNFI (Neath Port Talbot)
Aberhosan (Powys)
ABERKENFIG (Bridgend)
Aberllynfi (Powys)
ABERMORDDU (Flintshire)
Abermule (Powys)
Abernant (Carmarthenshire)
ABERSYCHAN (Torfaen)
Aberthin (Vale of Glamorgan)
ABERTILLERY (Blaenau Gwent)
ABERTRIDWR (Caerphilly)
Abertridwr (Powys)
ABERTYSSWG (Caerphilly)
Aberyscir (Powys)
Ablington (Gloucestershire)
Abney (Derbyshire)
ABOVE CHURCH (Staffordshire)
ABRAM (Greater Manchester)
Abson (Gloucestershire)

Aby (Lincolnshire)
ACASTER MALBIS (North Yorkshire)
ACASTER SELBY (North Yorkshire)
ACCRINGTON (Lancashire)
Acklam (North Yorkshire)
Ackleton (Shropshire)
ACKLINGTON (Northumberland)
ACKTON (West Yorkshire)
ACKWORTH MOOR TOP (West Yorkshire)
Acock's Green (West Midlands)
Acol (Kent)
Acomb (North Yorkshire)
ACOMB (Northumberland)
Acombe (Somerset)
Aconbury (Hereford & Worcester)
ACRE (Lancashire)
ACREFAIR (Wrexham)
ACRESFORD (Derbyshire)
Acton (Cheshire)
Acton (Hereford & Worcester)
Acton (Shropshire)
ACTON (Staffordshire)
Acton Beauchamp (Hereford & Worcester)
Acton Bridge (Cheshire)
ACTON BURNELL (Shropshire)
Acton Green (Hereford & Worcester)
ACTON PARK (Wrexham)
Acton Pigott (Shropshire)
Acton Round (Shropshire)
Acton Scott (Shropshire)
Acton Trussell (Staffordshire)
Acton Turville (Gloucestershire)
Adbaston (Staffordshire)
ADBOLTON (Nottinghamshire)
Adderley (Shropshire)
ADDERSTONE (Northumberland)
ADDINGHAM (West Yorkshire)
Addington (Kent)
Addlethorpe (Lincolnshire)
Adeney (Shropshire)
Adfa (Powys)
Adforton (Hereford & Worcester)
ADISHAM (Kent)
Adlestrop (Gloucestershire)
Adlingfleet (East Riding of Yorkshire)
ADLINGTON (Cheshire)
ADLINGTON (Lancashire)
ADMASTON (Shropshire)
Admaston (Staffordshire)
Admington (Warwickshire)
Adsborough (Somerset)
Adscombe (Somerset)
Adswood (Greater Manchester)
ADWALTON (West Yorkshire)
ADWICK LE STREET (South Yorkshire)
ADWICK UPON DEARNE (South Yorkshire)
AFFETSIDE (Greater Manchester)
Afon-wen (Flintshire)
AGGLETHORPE (North Yorkshire)
Aigburth (Merseyside)
Aike (East Riding of Yorkshire)
Aiketgate (Cumbria)
AIKHEAD (Cumbria)
Aikton (Cumbria)
Ailby (Lincolnshire)
Ailey (Hereford & Worcester)
Ainderby Quernhow (North Yorkshire)
Ainderby Steeple (North Yorkshire)
Ainsdale (Merseyside)
Ainsdale-on-Sea (Merseyside)
Ainstable (Cumbria)
AINSWORTH (Greater Manchester)
AINTHORPE (North Yorkshire)
Aintree (Merseyside)
AIRMYN (East Riding of Yorkshire)
Airton (North Yorkshire)
Aisby (Lincolnshire)

Aisgill (Cumbria)
Aisholt (Somerset)
Aiskew (North Yorkshire)
Aislaby (Durham)
Aislaby (North Yorkshire)
AISLABY MOOR (North Yorkshire)
Aisthorpe (Lincolnshire)
Akeld (Northumberland)
Alberbury (Shropshire)
Albrighton (Shropshire)
Alcaston (Shropshire)
Alcester (Warwickshire)
Alcester Lane End (West Midlands)
Aldborough (North Yorkshire)
Aldbrough (East Riding of Yorkshire)
Aldcliffe (Lancashire)
Alder Moor (Staffordshire)
ALDERCAR (Derbyshire)
Alderley (Gloucestershire)
Alderley Edge (Cheshire)
ALDERMANS GREEN (West Midlands)
Alderminster (Warwickshire)
Alderton (Gloucestershire)
Alderton (Shropshire)
ALDERWASLEY (Derbyshire)
Aldfield (North Yorkshire)
Aldford (Cheshire)
Aldingham (Cumbria)
Aldington (Hereford & Worcester)
Aldington (Kent)
Aldington Corner (Kent)
Aldon (Shropshire)
Aldoth (Cumbria)
ALDRIDGE (West Midlands)
Aldro (North Yorkshire)
Aldsworth (Gloucestershire)
Aldwark (Derbyshire)
Aldwark (North Yorkshire)
Aley (Somerset)
Alford (Lincolnshire)
Alford (Somerset)
ALFRETON (Derbyshire)
Alfrick (Hereford & Worcester)
Alfrick Pound (Hereford & Worcester)
Algarkirk (Lincolnshire)
Alhampton (Somerset)
Alkborough (Lincolnshire)
Alkerton (Gloucestershire)
Alkham (Kent)
Alkington (Shropshire)
Alkmonton (Derbyshire)
ALL STRETTON (Shropshire)
ALLASTON (Gloucestershire)
Allen End (Warwickshire)
Allendale (Northumberland)
ALLENHEADS (Northumberland)
Allensford (Durham)
Allensmore (Hereford & Worcester)
Allenton (Derbyshire)
Aller (Somerset)
ALLERBY (Cumbria)
Allerford (Somerset)
Allerston (North Yorkshire)
Allerthorpe (East Riding of Yorkshire)
Allerton (Merseyside)
ALLERTON (West Yorkshire)
ALLERTON BYWATER (West Yorkshire)
Allerton Mauleverer (North Yorkshire)
ALLESLEY (West Midlands)
Allestree (Derbyshire)
Allexton (Leicestershire)
ALLGREAVE (Cheshire)
Allhallows (Kent)
Allhallows-on-Sea (Kent)
Allimore Green (Staffordshire)
Allington (Kent)
Allington (Lincolnshire)

See paras. 2 and 4 of the User Guide 2003 if you can't find your place name.

Allithwaite (Cumbria)
ALLONBY (Cumbria)
Allostock (Cheshire)
Allowenshay (Somerset)
Allscott (Shropshire)
ALLTAMI (Flintshire)
Alltmawr (Powys)
Alltwalis (Carmarthenshire)
ALLTWEN (Neath Port Talbot)
Almeley (Hereford & Worcester)
Almeley Wooton (Hereford & Worcester)
ALMHOLME (South Yorkshire)
Almington (Staffordshire)
ALMONDBURY (West Yorkshire)
ALMONDSBURY (Gloucestershire)
Alne (North Yorkshire)
Alnham (Northumberland)
Alnmouth (Northumberland)
ALNWICK (Northumberland)
Alport (Derbyshire)
Alpraham (Cheshire)
ALREWAS (Staffordshire)
ALSAGER (Cheshire)
ALSAGERS BANK (Staffordshire)
Alsop en le Dale (Derbyshire)
ALSTON (Cumbria)
Alston Sutton (Somerset)
Alstone (Gloucestershire)
Alstone (Somerset)
Alstone Green (Staffordshire)
Alstonefield (Staffordshire)
ALT (Greater Manchester)
ALTHAM (Lancashire)
Althorpe (Lincolnshire)
ALTOFTS (West Yorkshire)
ALTON (Derbyshire)
Alton (Staffordshire)
Altrincham (Greater Manchester)
Alvanley (Cheshire)
Alvaston (Derbyshire)
Alvechurch (Hereford & Worcester)
ALVECOTE (Warwickshire)
ALVELEY (Shropshire)
ALVERTHORPE (West Yorkshire)
Alverton (Nottinghamshire)
Alveston (Gloucestershire)
Alveston (Warwickshire)
Alvingham (Lincolnshire)
Alvington (Gloucestershire)
Alwinton (Northumberland)
ALWOODLEY (West Yorkshire)
ALWOODLEY GATES (West Yorkshire)
Amber Hill (Lincolnshire)
AMBER ROW (Derbyshire)
AMBERGATE (Derbyshire)
Amberley (Gloucestershire)
AMBLE (Northumberland)
AMBLECOTE (West Midlands)
AMBLER THORN (West Yorkshire)
Ambleside (Cumbria)
Ambleston (Pembrokeshire)
Amcotts (Lincolnshire)
Amerton (Staffordshire)
Amington (Staffordshire)
Amlwch (Anglesey)
Amlwch Port (Anglesey)
AMMANFORD (Carmarthenshire)
Amotherby (North Yorkshire)
AMPLEFORTH (North Yorkshire)
Ampney Crucis (Gloucestershire)
Ampney St. Mary (Gloucestershire)
Ampney St. Peter (Gloucestershire)
AMROTH (Pembrokeshire)
Ancaster (Lincolnshire)
Anchor (Shropshire)
ANCROFT (Northumberland)
Anderby (Lincolnshire)
Andersea (Somerset)
Andersfield (Somerset)
Anderton (Cheshire)
Andoversford (Gloucestershire)
Anfield (Merseyside)
ANGELBANK (Shropshire)
Angersleigh (Somerset)
Angerton (Cumbria)
Angle (Pembrokeshire)
ANGRAM (North Yorkshire)
ANICK (Northumberland)
ANKLE HILL (Leicestershire)
Anlaby (East Riding of Yorkshire)
Annaside (Cumbria)

ANNESLEY (Nottinghamshire)
ANNESLEY WOODHOUSE (Nottinghamshire)
ANNFIELD PLAIN (Durham)
ANNITSFORD (Tyne & Wear)
ANNSCROFT (Shropshire)
Ansdell (Lancashire)
Ansford (Somerset)
ANSLEY (Warwickshire)
Anslow (Staffordshire)
Anslow Gate (Staffordshire)
Anslow Lees (Staffordshire)
Anstey (Leicestershire)
ANSTY (Warwickshire)
Anthorn (Cumbria)
Antrobus (Cheshire)
Anvil Green (Kent)
Anwick (Lincolnshire)
Apes Dale (Hereford & Worcester)
Apeton (Staffordshire)
Apley (Lincolnshire)
APPERKNOWLE (Derbyshire)
Apperley (Gloucestershire)
APPERLEY BRIDGE (West Yorkshire)
APPERLEY DENE (Northumberland)
APPERSETT (North Yorkshire)
Appleby (Lincolnshire)
Appleby Magna (Leicestershire)
Appleby Parva (Leicestershire)
Appleby-in-Westmorland (Cumbria)
Appledore (Kent)
Appledore Heath (Kent)
APPLEHAIGH (South Yorkshire)
Applethwaite (Cumbria)
APPLETON (Cheshire)
APPLETON ROEBUCK (North Yorkshire)
Appleton Thorn (Cheshire)
Appleton Wiske (North Yorkshire)
Appleton-le-Moors (North Yorkshire)
Appleton-le-Street (North Yorkshire)
APPLETREEWICK (North Yorkshire)
Appley (Somerset)
APPLEY BRIDGE (Lancashire)
Apsley Heath (Warwickshire)
ARBOURTHORNE (South Yorkshire)
Arcadia (Kent)
ARCHDDU (Carmarthenshire)
Archdeacon Newton (Durham)
Arclid Green (Cheshire)
Arddleen (Powys)
Ardens Grafton (Warwickshire)
ARDSLEY (South Yorkshire)
ARDSLEY EAST (West Yorkshire)
ARDWICK (Greater Manchester)
ARELEY KINGS (Hereford & Worcester)
ARGOED (Caerphilly)
Argoed (Shropshire)
Argoed Mill (Powys)
Arkendale (North Yorkshire)
Arkholme (Lancashire)
Arkle Town (North Yorkshire)
ARKLEBY (Cumbria)
ARKSEY (South Yorkshire)
ARKWRIGHT (Derbyshire)
Arle (Gloucestershire)
ARLECDON (Cumbria)
Arlescote (Warwickshire)
ARLESTON (Shropshire)
Arley (Cheshire)
ARLEY (Warwickshire)
Arlingham (Gloucestershire)
Arlington (Gloucestershire)
Armaside (Cumbria)
Armathwaite (Cumbria)
ARMITAGE (Staffordshire)
ARMITAGE BRIDGE (West Yorkshire)
ARMLEY (West Yorkshire)
ARMSHEAD (Staffordshire)
ARMTHORPE (South Yorkshire)
Arnaby (Cumbria)
ARNCLIFFE (North Yorkshire)
ARNCLIFFE COTE (North Yorkshire)
Arnesby (Leicestershire)
Arnfield (Derbyshire)
ARNO'S VALE (Bristol)
Arnold (East Riding of Yorkshire)
ARNOLD (Nottinghamshire)
Arnside (Cumbria)
AROWRY (Wrexham)
Arrad Foot (Cumbria)
Arram (East Riding of Yorkshire)
ARRATHORNE (North Yorkshire)

Arrow (Warwickshire)
Arrowfield Top (Hereford & Worcester)
ARSCOTT (Shropshire)
Arthington (West Yorkshire)
ARTHURSDALE (West Yorkshire)
ASBY (Cumbria)
Ascott (Warwickshire)
Asenby (North Yorkshire)
ASFORDBY (Leicestershire)
ASFORDBY HILL (Leicestershire)
Asgarby (Lincolnshire)
ASH (Kent)
Ash (Somerset)
ASH GREEN (Warwickshire)
Ash Magna (Shropshire)
Ash Parva (Shropshire)
Ash Priors (Somerset)
Ashbourne (Derbyshire)
Ashbourne Green (Derbyshire)
Ashbrittle (Somerset)
Ashby (Lincolnshire)
Ashby by Partney (Lincolnshire)
Ashby cum Fenby (Lincolnshire)
Ashby de la Launde (Lincolnshire)
Ashby Folville (Leicestershire)
Ashby Magna (Leicestershire)
Ashby Parva (Leicestershire)
Ashby Puerorum (Lincolnshire)
ASHBY-DE-LA-ZOUCH (Leicestershire)
Ashchurch (Gloucestershire)
Ashcombe (Somerset)
Ashcott (Somerset)
Ashfield (Hereford & Worcester)
Ashfields (Shropshire)
Ashford (Kent)
Ashford Bowdler (Shropshire)
Ashford Carbonel (Shropshire)
Ashford in the Water (Derbyshire)
Ashill (Somerset)
ASHINGTON (Northumberland)
Ashington (Somerset)
Ashleworth (Gloucestershire)
Ashleworth Quay (Gloucestershire)
Ashley (Cheshire)
Ashley (Gloucestershire)
ASHLEY (Kent)
Ashley (Staffordshire)
ASHLEY DOWN (Bristol)
Ashley Moor (Hereford & Worcester)
Ashmead Green (Gloucestershire)
Ashorne (Warwickshire)
ASHOVER (Derbyshire)
ASHOVER HAY (Derbyshire)
Ashow (Warwickshire)
Ashperton (Hereford & Worcester)
Ashton (Cheshire)
Ashton (Hereford & Worcester)
Ashton (Somerset)
Ashton under Hill (Hereford & Worcester)
Ashton upon Mersey (Greater Manchester)
Ashton Watering (Somerset)
ASHTON-IN-MAKERFIELD (Greater Manchester)
ASHTON-UNDER-LYNE (Greater Manchester)
ASHTONGATE (Bristol)
Ashurst (Kent)
Ashwell (Somerset)
ASHWICK (Somerset)
Ashwood (Staffordshire)
Askam in Furness (Cumbria)
ASKE HALL (North Yorkshire)
ASKERN (South Yorkshire)
ASKHAM (Cumbria)
Askham (Nottinghamshire)
ASKHAM BRYAN (North Yorkshire)
Askham Richard (North Yorkshire)
ASKRIGG (North Yorkshire)
Askwith (North Yorkshire)
Aslackby (Lincolnshire)
Aslockton (Nottinghamshire)
Asney (Somerset)
ASPATRIA (Cumbria)
Asperton (Lincolnshire)
Aspley (Staffordshire)
ASPULL (Greater Manchester)
ASPULL COMMON (Greater Manchester)
Asselby (East Riding of Yorkshire)
Asserby (Lincolnshire)
Asserby Turn (Lincolnshire)
Astbury (Cheshire)
Asterby (Lincolnshire)
ASTERLEY (Shropshire)

See paras. 2 and 4 of the User Guide 2003 if you can't find your place name.

Asterton (Shropshire)
ASTLEY (Greater Manchester)
ASTLEY (Hereford & Worcester)
Astley (Shropshire)
ASTLEY (Warwickshire)
ASTLEY (West Yorkshire)
ASTLEY ABBOTS (Shropshire)
ASTLEY BRIDGE (Greater Manchester)
ASTLEY CROSS (Hereford & Worcester)
ASTLEY GREEN (Greater Manchester)
ASTLEY TOWN (Hereford & Worcester)
Aston (Cheshire)
Aston (Derbyshire)
ASTON (Flintshire)
Aston (Hereford & Worcester)
ASTON (Shropshire)
ASTON (South Yorkshire)
Aston (Staffordshire)
Aston (West Midlands)
ASTON BOTTERELL (Shropshire)
Aston Cantlow (Warwickshire)
Aston Crews (Hereford & Worcester)
Aston Cross (Gloucestershire)
Aston Fields (Hereford & Worcester)
Aston Flamville (Leicestershire)
Aston Heath (Cheshire)
Aston Ingham (Hereford & Worcester)
Aston juxta Mondrum (Cheshire)
Aston Magna (Gloucestershire)
Aston Munslow (Shropshire)
Aston on Clun (Shropshire)
Aston Pigott (Shropshire)
Aston Rogers (Shropshire)
Aston Somerville (Hereford & Worcester)
Aston Subedge (Gloucestershire)
Aston-Eyre (Shropshire)
Aston-upon-Trent (Derbyshire)
Astonlane (Shropshire)
ASTWITH (Derbyshire)
Astwood (Hereford & Worcester)
Astwood Bank (Hereford & Worcester)
Aswarby (Lincolnshire)
Aswardby (Lincolnshire)
Atch Lench (Hereford & Worcester)
ATCHAM (Shropshire)
Athelney (Somerset)
Atherstone (Somerset)
ATHERSTONE (Warwickshire)
Atherstone on Stour (Warwickshire)
ATHERTON (Greater Manchester)
Atley Hill (North Yorkshire)
Atlow (Derbyshire)
ATTENBOROUGH (Nottinghamshire)
Atterby (Lincolnshire)
ATTERCLIFFE (South Yorkshire)
Atterley (Shropshire)
Atterton (Leicestershire)
ATTLEBOROUGH (Warwickshire)
Atwick (East Riding of Yorkshire)
Auberrow (Hereford & Worcester)
Aubourn (Lincolnshire)
AUCKLEY (South Yorkshire)
AUDENSHAW (Greater Manchester)
Audlem (Cheshire)
AUDLEY (Staffordshire)
Audmore (Staffordshire)
AUDNAM (West Midlands)
Aughertree (Cumbria)
AUGHTON (East Riding of Yorkshire)
Aughton (Lancashire)
AUGHTON (South Yorkshire)
Aughton Park (Lancashire)
Aulden (Hereford & Worcester)
AULT HUCKNALL (Derbyshire)
Aunby (Lincolnshire)
Aunsby (Lincolnshire)
Aust (Gloucestershire)
Austendike (Lincolnshire)
AUSTERFIELD (South Yorkshire)
AUSTERLANDS (Greater Manchester)
AUSTHORPE (West Yorkshire)
Austonley (West Yorkshire)
AUSTREY (Warwickshire)
Austwick (North Yorkshire)
Authorpe (Lincolnshire)
Authorpe Row (Lincolnshire)
Avening (Gloucestershire)
Averham (Nottinghamshire)
Avon Dassett (Warwickshire)
AVONMOUTH (Bristol)
Awkley (Gloucestershire)

Awre (Gloucestershire)
AWSWORTH (Nottinghamshire)
Axborough (Hereford & Worcester)
Axbridge (Somerset)
Axton (Flintshire)
AYCLIFFE (Durham)
AYDON (Northumberland)
Aylburton (Gloucestershire)
AYLE (Cumbria)
Aylesby (Lincolnshire)
Aylesford (Kent)
AYLESHAM (Kent)
Aylestone (Leicestershire)
Aylestone Park (Leicestershire)
Aylton (Gloucestershire)
Aylworth (Gloucestershire)
Aymestrey (Hereford & Worcester)
AYSGARTH (North Yorkshire)
Ayside (Cumbria)
Azerley (North Yorkshire)

B

BABBINGTON (Nottinghamshire)
Babbinswood (Shropshire)
Babcary (Somerset)
Babel (Carmarthenshire)
Babell (Flintshire)
BABINGTON (Somerset)
BABWORTH (Nottinghamshire)
Bachau (Anglesey)
Bache (Shropshire)
Bacheldre (Powys)
Back o' th' Brook (Staffordshire)
Backbarrow (Cumbria)
Backe (Carmarthenshire)
Backford (Cheshire)
Backford Cross (Cheshire)
BACKWELL (Somerset)
BACKWORTH (Tyne & Wear)
Bacon's End (West Midlands)
Bacton (Hereford & Worcester)
BACUP (Lancashire)
BADDELEY EDGE (Staffordshire)
BADDELEY GREEN (Staffordshire)
Baddesley Clinton (Warwickshire)
BADDESLEY ENSOR (Warwickshire)
BADGER (Shropshire)
Badgers Mount (Kent)
Badgeworth (Gloucestershire)
Badgworth (Somerset)
Badlesmere (Kent)
Badsey (Hereford & Worcester)
BADSWORTH (West Yorkshire)
Bag Enderby (Lincolnshire)
Bagby (North Yorkshire)
Bagendon (Gloucestershire)
BAGGINSWOOD (Shropshire)
BAGGROW (Cumbria)
Bagham (Kent)
BAGILLT (Flintshire)
Baginton (Warwickshire)
BAGLAN (Neath Port Talbot)
Bagley (Shropshire)
Bagley (Somerset)
BAGLEY (West Yorkshire)
BAGNALL (Staffordshire)
BAGOT (Shropshire)
BAGSTONE (Gloucestershire)
BAGTHORPE (Nottinghamshire)
BAGWORTH (Leicestershire)
Bagwy Llydiart (Hereford & Worcester)
BAILDON (West Yorkshire)
BAILDON GREEN (West Yorkshire)
Baileyhead (Cumbria)
BAILIFF BRIDGE (West Yorkshire)
Bailrigg (Lancashire)
BAINBRIDGE (North Yorkshire)
Bainton (East Riding of Yorkshire)
Bakewell (Derbyshire)
BALBY (South Yorkshire)
Baldersby (North Yorkshire)
Baldersby St. James (North Yorkshire)
BALDERSTONE (Greater Manchester)
Balderstone (Lancashire)
Balderton (Nottinghamshire)
BALDWIN'S GATE (Staffordshire)
Baldwinholme (Cumbria)
Balk (North Yorkshire)
Balkholme (East Riding of Yorkshire)
BALL (Shropshire)

BALL GREEN (Staffordshire)
Ball Haye Green (Staffordshire)
Ball's Green (Gloucestershire)
BALLARDS GREEN (Warwickshire)
Ballidon (Derbyshire)
Ballingham (Hereford & Worcester)
Balmer Heath (Shropshire)
BALNE (North Yorkshire)
BALSALL (West Midlands)
BALSALL COMMON (West Midlands)
Balsall Heath (West Midlands)
BALSALL STREET (West Midlands)
Balterley (Staffordshire)
Balterley Green (Staffordshire)
Baltonsborough (Somerset)
Bamber Bridge (Lancashire)
Bamburgh (Northumberland)
Bamford (Derbyshire)
BAMFORD (Greater Manchester)
Bampton (Cumbria)
Bampton Grange (Cumbria)
Banc-y-ffordd (Carmarthenshire)
BANCFFOSFELEM (Carmarthenshire)
Bancycapel (Carmarthenshire)
Bancyfelin (Carmarthenshire)
Bandrake Head (Cumbria)
Bangor's Green (Lancashire)
BANGOR-IS-Y-COED (Wrexham)
Bank Ground (Cumbria)
Bank Newton (North Yorkshire)
Bank Street (Hereford & Worcester)
BANK TOP (Lancashire)
BANK TOP (West Yorkshire)
BANKS (Cumbria)
Banks (Lancashire)
Banks Green (Hereford & Worcester)
Banwell (Somerset)
BAPTIST MILLS (Bristol)
Bapchild (Kent)
Barber Booth (Derbyshire)
Barber Green (Cumbria)
BARBON (Cumbria)
Barbridge (Cheshire)
Barcheston (Warwickshire)
Barclose (Cumbria)
BARCROFT (West Yorkshire)
BARDEN (North Yorkshire)
Barden Park (Kent)
Bardney (Lincolnshire)
BARDON (Leicestershire)
BARDON MILL (Northumberland)
Bardsea (Cumbria)
BARDSEY (West Yorkshire)
BARDSEY (Greater Manchester)
Bare (Lancashire)
Barewood (Hereford & Worcester)
Barford (Warwickshire)
BARFRESTONE (Kent)
BARGATE (Derbyshire)
BARGOED (Caerphilly)
BARHAM (Kent)
Barholm (Lincolnshire)
Barkby (Leicestershire)
Barkby Thorpe (Leicestershire)
Barkers Green (Shropshire)
Barkestone-le-Vale (Leicestershire)
BARKISLAND (West Yorkshire)
Barkston (Lincolnshire)
Barkston Ash (North Yorkshire)
BARLASTON (Staffordshire)
BARLBOROUGH (Derbyshire)
BARLBY (North Yorkshire)
BARLESTONE (Leicestershire)
Barley (Lancashire)
BARLEY HOLE (South Yorkshire)
Barlings (Lincolnshire)
BARLOW (Derbyshire)
BARLOW (North Yorkshire)
BARLOW (Tyne & Wear)
Barmby Moor (East Riding of Yorkshire)
BARMBY ON THE MARSH (East Riding of Yorkshire)
Barming Heath (Kent)
Barmpton (Durham)
Barmston (East Riding of Yorkshire)
BARNACLE (Warwickshire)
Barnard Castle (Durham)
BARNBURGH (South Yorkshire)
BARNBY DUN (South Yorkshire)
Barnby in the Willows (Nottinghamshire)
BARNBY MOOR (Nottinghamshire)
Barnes Street (Kent)

Barnetby le Wold (Lincolnshire)
Barnhill (Cheshire)
Barningham (Durham)
Barnoldby le Beck (Lincolnshire)
Barnoldswick (Lancashire)
BARNSDALE BAR (North Yorkshire)
Barnsley (Gloucestershire)
Barnsley (Shropshire)
BARNSLEY (South Yorkshire)
Barnsole (Kent)
Barnston (Merseyside)
BARNSTONE (Nottinghamshire)
Barnt Green (Hereford & Worcester)
Barnton (Cheshire)
Barnwood (Gloucestershire)
Baron's Cross (Hereford & Worcester)
Baronwood (Cumbria)
BARRAS (Cumbria)
Barrasford (Northumberland)
Barrets Green (Cheshire)
Barrington (Somerset)
Barrow (Gloucestershire)
Barrow (Lancashire)
BARROW (Shropshire)
Barrow (Somerset)
BARROW BRIDGE (Greater Manchester)
Barrow Burn (Northumberland)
Barrow Gurney (Somerset)
Barrow Haven (Lincolnshire)
BARROW HILL (Derbyshire)
Barrow Island (Cumbria)
BARROW NOOK (Lancashire)
Barrow upon Soar (Leicestershire)
Barrow upon Trent (Derbyshire)
BARROW VALE (Somerset)
BARROW'S GREEN (Cheshire)
Barrow-in-Furness (Cumbria)
Barrow-upon-Humber (Lincolnshire)
Barrowby (Lincolnshire)
BARROWFORD (Lancashire)
Barry (Vale of Glamorgan)
Barry Island (Vale of Glamorgan)
Barsby (Leicestershire)
Barston (West Midlands)
Bartestree (Hereford & Worcester)
BARTHOMLEY (Cheshire)
Bartley Green (West Midlands)
Barton (Cheshire)
Barton (Cumbria)
Barton (Gloucestershire)
Barton (Hereford & Worcester)
Barton (Lancashire)
Barton (North Yorkshire)
Barton (Warwickshire)
Barton End (Gloucestershire)
BARTON GREEN (Staffordshire)
Barton in Fabis (Nottinghamshire)
Barton in the Beans (Leicestershire)
Barton St. David (Somerset)
BARTON UPON IRWELL (Greater Manchester)
Barton Waterside (Lincolnshire)
Barton-le-Street (North Yorkshire)
Barton-le-Willows (North Yorkshire)
Barton-on-the-Heath (Warwickshire)
Barton-under-Needwood (Staffordshire)
Barton-upon-Humber (Lincolnshire)
BARTON HILL (Bristol)
BARUGH (South Yorkshire)
BARUGH GREEN (South Yorkshire)
Barwell (Leicestershire)
Barwick (Somerset)
BARWICK IN ELMET (West Yorkshire)
Baschurch (Shropshire)
Bascote (Warwickshire)
Bascote Heath (Warwickshire)
BASFORD GREEN (Staffordshire)
Bashall Eaves (Lancashire)
Bashall Town (Lancashire)
BASLOW (Derbyshire)
Bason Bridge (Somerset)
Bassaleg (Newport)
Bassenthwaite (Cumbria)
BASSINGFIELD (Nottinghamshire)
Bassingham (Lincolnshire)
Bassingthorpe (Lincolnshire)
Basted (Kent)
Baston (Lincolnshire)
Batch (Somerset)
Batcombe (Somerset)
Bate Heath (Cheshire)
Bath (Somerset)

Bathampton (Somerset)
Bathealton (Somerset)
Batheaston (Somerset)
Bathford (Somerset)
BATHLEY (Nottinghamshire)
Bathpool (Somerset)
Bathway (Somerset)
BATLEY (West Yorkshire)
Batsford (Gloucestershire)
BATTERSBY (North Yorkshire)
Battle (Powys)
Battleborough (Somerset)
Battledown (Gloucestershire)
Battlefield (Shropshire)
Battleton (Somerset)
BATTYE FORD (West Yorkshire)
Baughton (Hereford & Worcester)
Baumber (Lincolnshire)
Baunton (Gloucestershire)
BAVENEY WOOD (Shropshire)
Bawdrip (Somerset)
BAWTRY (South Yorkshire)
BAXENDEN (Lancashire)
BAXTERLEY (Warwickshire)
Baxton (Kent)
Bay Horse (Lancashire)
Baybridge (Northumberland)
Baycliff (Cumbria)
Bayford (Somerset)
Bayley's Hill (Kent)
Baysdale Abbey (North Yorkshire)
BAYSDALE MOOR (North Yorkshire)
Baysham (Hereford & Worcester)
BAYSTON HILL (Shropshire)
BAYTON (Hereford & Worcester)
BAYTON COMMON (Hereford & Worcester)
Beach (Gloucestershire)
Beachborough (Kent)
Beachley (Gloucestershire)
Beacon Hill (Kent)
Beacon Hill (Nottinghamshire)
Beadlam (North Yorkshire)
BEADNELL (Northumberland)
BEAL (North Yorkshire)
Beal (Northumberland)
Beam Hill (Staffordshire)
Beamhurst (Staffordshire)
BEAMISH (Durham)
BEAMSLEY (North Yorkshire)
Bean (Kent)
Beanley (Northumberland)
BEARDWOOD (Lancashire)
Bearley (Warwickshire)
Bearley Cross (Warwickshire)
BEARPARK (Durham)
Bearstead (Kent)
Bearstone (Shropshire)
BEARWOOD (West Midlands)
BEAUCHIEF (South Yorkshire)
Beaudesert (Warwickshire)
BEAUFORT (Blaenau Gwent)
Beaumaris (Anglesey)
Beaumont (Cumbria)
Beaumont Hill (Durham)
Beausale (Warwickshire)
Beaver (Kent)
Beaver Green (Kent)
Bebington (Merseyside)
BEBSIDE (Northumberland)
Becconsall (Lancashire)
Beck Foot (Cumbria)
BECK HOLE (North Yorkshire)
Beck Side (Cumbria)
BECKBURY (Shropshire)
Beckering (Lincolnshire)
Beckermet (Cumbria)
Beckfoot (Cumbria)
Beckford (Hereford & Worcester)
Beckingham (Lincolnshire)
BECKINGHAM (Nottinghamshire)
Beckington (Somerset)
Beckjay (Shropshire)
Becks (West Yorkshire)
Beckside (Cumbria)
Beckwithshaw (North Yorkshire)
Bedale (North Yorkshire)
BEDBURN (Durham)
BEDDAU (Rhondda Cynon Taff)
Bedgebury Cross (Kent)
Bedlam (North Yorkshire)
Bedlam Lane (Kent)

BEDLINGTON (Northumberland)
BEDLINOG (Merthyr Tydfil)
BEDMINSTER (Bristol)
BEDMINSTER DOWN (Bristol)
BEDNALL (Staffordshire)
Bedstone (Shropshire)
BEDWAS (Caerphilly)
BEDWELLTY (Caerphilly)
BEDWORTH (Warwickshire)
BEDWORTH WOODLANDS (Warwickshire)
Beeby (Leicestershire)
BEECH (Staffordshire)
Beeford (East Riding of Yorkshire)
Beeley (Derbyshire)
Beelsby (Lincolnshire)
Beer (Somerset)
Beercrocombe (Somerset)
Beesby (Lincolnshire)
Beeston (Cheshire)
BEESTON (Nottinghamshire)
BEESTON (West Yorkshire)
Beetham (Cumbria)
Beetham (Somerset)
Began (Cardiff)
BEGELLY (Pembrokeshire)
Beggar's Bush (Powys)
BEGGARINTON HILL (West Yorkshire)
Beguildy (Powys)
BEIGHTON (South Yorkshire)
Beighton Hill (Derbyshire)
Bekesbourne (Kent)
Bekesbourne Hill (Kent)
Belbroughton (Hereford & Worcester)
Belchford (Lincolnshire)
BELFORD (Northumberland)
Belgrave (Leicestershire)
Bell Busk (North Yorkshire)
Bell End (Hereford & Worcester)
Bell Heath (Hereford & Worcester)
Bell o' th'Hill (Cheshire)
Bellasize (East Riding of Yorkshire)
Belle Vale (Merseyside)
Belle Vue (Cumbria)
BELLE VUE (West Yorkshire)
Belleau (Lincolnshire)
BELLERBY (North Yorkshire)
Bellimoor (Hereford & Worcester)
BELLINGHAM (Northumberland)
BELLSHILL (Northumberland)
BELLUTON (Somerset)
BELMONT (Lancashire)
BELPER (Derbyshire)
BELPER LANE END (Derbyshire)
BELPH (Derbyshire)
BELSAY (Northumberland)
Belsay Castle (Northumberland)
BELTHORN (Lancashire)
Beltinge (Kent)
BELTINGHAM (Northumberland)
Beltoft (Lincolnshire)
Belton (East Riding of Yorkshire)
Belton (Leicestershire)
Belton (Lincolnshire)
Beltring (Kent)
Belvoir (Leicestershire)
BEMERSLEY GREEN (Staffordshire)
Bempton (East Riding of Yorkshire)
BEN RHYDDING (West Yorkshire)
Benenden (Kent)
BENFIELDSIDE (Durham)
Bengeworth (Hereford & Worcester)
Beningbrough (North Yorkshire)
Benington (Lincolnshire)
Benllech (Anglesey)
Bennet Head (Cumbria)
Bennetland (East Riding of Yorkshire)
Bennington Sea End (Lincolnshire)
Benniworth (Lincolnshire)
Benover (Kent)
BENTHALL (Shropshire)
Bentham (Gloucestershire)
Bentlawn (Shropshire)
Bentley (East Riding of Yorkshire)
BENTLEY (South Yorkshire)
BENTLEY (Warwickshire)
Bentley Heath (West Midlands)
BENTLEY RISE (South Yorkshire)
Beoley (Hereford & Worcester)
Berea (Pembrokeshire)
Berhill (Somerset)
Berkeley (Gloucestershire)

See paras. 2 and 4 of the User Guide 2003 if you can't find your place name.

Berkeley Heath (Gloucestershire)
Berkeley Road (Gloucestershire)
Berkley (Somerset)
BERKSWELL (West Midlands)
Berrier (Cumbria)
Berriew (Powys)
Berrington (Hereford & Worcester)
BERRINGTON (Northumberland)
Berrington (Shropshire)
Berrington Green (Hereford & Worcester)
Berrow (Hereford & Worcester)
Berrow (Somerset)
Berrow Green (Hereford & Worcester)
BERRY BROW (West Yorkshire)
BERRY HILL (Gloucestershire)
Berry Hill (Pembrokeshire)
BERSHAM (Wrexham)
Berthengam (Flintshire)
Berwick Hill (Northumberland)
BERWICK-UPON-TWEED (Northumberland)
Bescaby (Leicestershire)
Bescar (Cumbria)
Besford (Hereford & Worcester)
Besford (Shropshire)
BESOM HILL (Greater Manchester)
BESSACARR (South Yorkshire)
BESSES O' TH' BARN (Greater Manchester)
Bessingby (East Riding of Yorkshire)
BESTHORPE (Nottinghamshire)
Beswick (East Riding of Yorkshire)
Betchcott (Shropshire)
Bethel (Anglesey)
Bethel (Powys)
Bethersden (Kent)
Bethesda (Pembrokeshire)
Bethlehem (Carmarthenshire)
Betley (Staffordshire)
Betsham (Kent)
BETTESHANGER (Kent)
Bettisfield (Wrexham)
Betton (Shropshire)
BETTON STRANGE (Shropshire)
Bettws (Newport)
Bettws Cedewain (Powys)
Bettws-Newydd (Monmouthshire)
BETWS (Bridgend)
BETWS (Carmarthenshire)
Betws Gwerfil Goch (Denbighshire)
Beulah (Powys)
BEVERCOTES (Nottinghamshire)
Beverley (East Riding of Yorkshire)
Beverstone (Gloucestershire)
Bevington (Gloucestershire)
Bewaldeth (Cumbria)
Bewcastle (Cumbria)
BEWDLEY (Hereford & Worcester)
BEWERLEY (North Yorkshire)
Bewholme (East Riding of Yorkshire)
Bewlbridge (Kent)
Bibstone (Gloucestershire)
Bibury (Gloucestershire)
Bickenhill (West Midlands)
Bicker (Lincolnshire)
Bicker Bar (Lincolnshire)
Bicker Gauntlet (Lincolnshire)
BICKERSHAW (Greater Manchester)
BICKERSTAFFE (Lancashire)
Bickerton (Cheshire)
Bickerton (North Yorkshire)
BICKERTON (Northumberland)
Bickford (Staffordshire)
Bickley (Cheshire)
Bickley (Hereford & Worcester)
Bickley (North Yorkshire)
Bickley Moss (Cheshire)
Bicknoller (Somerset)
Bicknor (Kent)
Bicton (Hereford & Worcester)
Bicton (Shropshire)
Bidborough (Kent)
Biddenden (Kent)
Biddenden Green (Kent)
Biddisham (Somerset)
Biddlestone (Northumberland)
BIDDULPH (Staffordshire)
BIDDULPH MOOR (Staffordshire)
BIDEFORD (Devon)
Bidford-on-Avon (Warwickshire)
Bidston (Merseyside)
Bielby (East Riding of Yorkshire)
Bigby (Lincolnshire)

Biggar (Cumbria)
Biggin (Derbyshire)
BIGGIN (North Yorkshire)
Bigland Hall (Cumbria)
Biglands (Cumbria)
BIGRIGG (Cumbria)
BILBOROUGH (Nottinghamshire)
Bilbrook (Somerset)
Bilbrook (Staffordshire)
Bilbrough (North Yorkshire)
BILDERSHAW (Durham)
Billesdon (Leicestershire)
Billesley (Warwickshire)
Billingborough (Lincolnshire)
BILLINGE (Merseyside)
Billingham (Durham)
Billinghay (Lincolnshire)
BILLINGLEY (South Yorkshire)
BILLINGSLEY (Shropshire)
BILLINGTON (Lancashire)
Billington (Staffordshire)
BILLY ROW (Durham)
Bilsborrow (Lancashire)
Bilsby (Lincolnshire)
Bilsington (Kent)
BILSTHORPE (Nottinghamshire)
BILSTHORPE MOOR (Nottinghamshire)
BILSTON (West Midlands)
Bilstone (Leicestershire)
Bilting (Kent)
Bilton (East Riding of Yorkshire)
Bilton (North Yorkshire)
BILTON (Northumberland)
Bilton (Warwickshire)
BILTON BANKS (Northumberland)
BILTON DENE (North Yorkshire)
Binbrook (Lincolnshire)
BINCHESTER BLOCKS (Durham)
BINEGAR (Somerset)
BINGFIELD (Northumberland)
BINGHAM (Nottinghamshire)
BINGLEY (West Yorkshire)
Bings (Shropshire)
BINLEY (West Midlands)
Binton (Warwickshire)
Binweston (Shropshire)
BIRCH (Greater Manchester)
Birch Cross (Staffordshire)
Birch Heath (Cheshire)
Birch Hill (Cheshire)
BIRCH VALE (Derbyshire)
Birch Wood (Somerset)
BIRCHENCLIFFE (West Yorkshire)
Bircher (Hereford & Worcester)
BIRCHFIELD (West Midlands)
Birchgrove (Cardiff)
Birchington (Kent)
BIRCHLEY HEATH (Warwickshire)
Birchover (Derbyshire)
Birchyfield (Hereford & Worcester)
BIRCOTES (Nottinghamshire)
Bird End (West Midlands)
BIRDFORTH (North Yorkshire)
Birdingbury (Warwickshire)
Birdlip (Gloucestershire)
BIRDOSWALD (Cumbria)
BIRDS EDGE (West Yorkshire)
Birdsall (North Yorkshire)
BIRDSGREEN (Shropshire)
BIRDWELL (South Yorkshire)
Birdwood (Gloucestershire)
BIRKACRE (Lancashire)
Birkby (North Yorkshire)
Birkdale (Merseyside)
Birkenhead (Merseyside)
BIRKENSHAW (West Yorkshire)
Birkholme (Lincolnshire)
BIRKIN (North Yorkshire)
BIRKS (West Yorkshire)
BIRKSHAW (Northumberland)
Birley (Hereford & Worcester)
BIRLEY CARR (South Yorkshire)
Birling (Kent)
BIRLING (Northumberland)
Birlingham (Hereford & Worcester)
BIRMINGHAM (West Midlands)
Birstall (Leicestershire)
BIRSTALL (West Yorkshire)
BIRSTWITH (North Yorkshire)
Birthorpe (Lincolnshire)
Birtley (Hereford & Worcester)

Birtley (Northumberland)
BIRTLEY (Tyne & Wear)
Birts Street (Hereford & Worcester)
Biscathorpe (Lincolnshire)
Bishampton (Hereford & Worcester)
BISHOP AUCKLAND (Durham)
Bishop Burton (East Riding of Yorkshire)
BISHOP MIDDLEHAM (Durham)
Bishop Monkton (North Yorkshire)
Bishop Norton (Lincolnshire)
BISHOP SUTTON (Somerset)
Bishop Thornton (North Yorkshire)
Bishop Wilton (East Riding of Yorkshire)
Bishop's Castle (Shropshire)
Bishop's Cleeve (Gloucestershire)
Bishop's Frome (Hereford & Worcester)
Bishop's Itchington (Warwickshire)
Bishop's Norton (Gloucestershire)
Bishop's Offley (Staffordshire)
Bishop's Tachbrook (Warwickshire)
Bishop's Wood (Staffordshire)
Bishopbridge (Lincolnshire)
Bishops Hull (Somerset)
Bishops Lydeard (Somerset)
Bishopsbourne (Kent)
BISHOPSTON (Swansea)
BISHOPSTON (Bristol)
Bishopstone (Hereford & Worcester)
Bishopstone (Kent)
Bishopswood (Somerset)
BISHOPSWORTH (Bristol)
BISHOPTHORPE (North Yorkshire)
Bishopton (Durham)
Bishopton (Warwickshire)
Bishton (Newport)
Bishton (Staffordshire)
Bisley (Gloucestershire)
Bispham (Lancashire)
BISPHAM GREEN (Lancashire)
Bitchet Green (Kent)
Bitchfield (Lincolnshire)
BITTERLEY (Shropshire)
Bitteswell (Leicestershire)
BITTON (Gloucestershire)
Blaby (Leicestershire)
BLACK CALLERTON (Tyne & Wear)
Black Heddon (Northumberland)
BLACK LANE (Greater Manchester)
BLACK LANE ENDS (Lancashire)
BLACK MOOR (West Yorkshire)
Black Tar (Pembrokeshire)
BLACKBANK (Warwickshire)
Blackbeck (Cumbria)
Blackbrook (Derbyshire)
Blackbrook (Staffordshire)
BLACKBURN (Lancashire)
Blackden Heath (Cheshire)
Blackdyke (Cumbria)
BLACKENALL HEATH (West Midlands)
BLACKER (South Yorkshire)
BLACKER HILL (South Yorkshire)
Blackford (Cumbria)
Blackford (Somerset)
BLACKFORD BRIDGE (Greater Manchester)
BLACKFORDBY (Leicestershire)
BLACKHALL (Durham)
BLACKHALL COLLIERY (Durham)
BLACKHEATH (West Midlands)
BLACKHILL (Durham)
Blackjack (Lincolnshire)
Blackland (Somerset)
BLACKLEY (Greater Manchester)
Blackmarstone (Hereford & Worcester)
BLACKMILL (Bridgend)
Blackmoor (Somerset)
Blackmoorfoot (West Yorkshire)
BLACKO (Lancashire)
BLACKPILL (Swansea)
Blackpool (Lancashire)
Blackpool Gate (Cumbria)
BLACKROCK (Monmouthshire)
BLACKROD (Greater Manchester)
Blackshaw Head (West Yorkshire)
BLACKSNAPE (Lancashire)
Blacktoft (East Riding of Yorkshire)
Blackwall (Derbyshire)
Blackwater (Somerset)
Blackwell (Cumbria)
BLACKWELL (Derbyshire)
Blackwell (Durham)
Blackwell (Hereford & Worcester)

See paras. 2 and 4 of the User Guide 2003 if you can't find your place name.

Blackwell (Warwickshire)
Blackwellsend Green (Gloucestershire)
BLACKWOOD (Caerphilly)
Blackwood Hill (Staffordshire)
Blacon (Cheshire)
Bladbean (Kent)
Bladon (Somerset)
Blaen Dyryn (Powys)
Blaen-y-Coed (Carmarthenshire)
BLAEN-Y-CWM (Blaenau Gwent)
BLAENAVON (Torfaen)
Blaenffos (Pembrokeshire)
BLAENGARW (Bridgend)
BLAENGWRACH (Neath Port Talbot)
BLAENGWYNFI (Neath Port Talbot)
BLAENLLECHAU (Rhondda Cynon Taff)
BLAENRHONDDA (Rhondda Cynon Taff)
Blaenwaun (Carmarthenshire)
Blagdon (Somerset)
Blagdon Hill (Somerset)
BLAGILL (Cumbria)
BLAGUEGATE (Lancashire)
BLAINA (Blaenau Gwent)
Blaisdon (Gloucestershire)
Blakebrook (Hereford & Worcester)
Blakedown (Hereford & Worcester)
BLAKELEY LANE (Staffordshire)
Blakemere (Cheshire)
Blakemere (Hereford & Worcester)
BLAKENEY (Gloucestershire)
Blakenhall (Cheshire)
BLAKENHALL (West Midlands)
Blakeshall (Hereford & Worcester)
Blanchland (Northumberland)
Bland Hill (North Yorkshire)
Blankney (Lincolnshire)
Blaston (Leicestershire)
Blawith (Cumbria)
BLAXTON (South Yorkshire)
BLAYDON (Tyne & Wear)
Bleadney (Somerset)
Bleadon (Somerset)
Bleak Street (Somerset)
Blean (Kent)
Bleasby (Nottinghamshire)
Bleasdale (Lancashire)
Bleatarn (Cumbria)
Bleathwood (Hereford & Worcester)
Bleddfa (Powys)
Bledington (Gloucestershire)
Blencarn (Cumbria)
BLENCOGO (Cumbria)
BLENNERHASSET (Cumbria)
Bletchley (Shropshire)
Bletherston (Pembrokeshire)
BLIDWORTH (Nottinghamshire)
BLIDWORTH BOTTOMS (Nottinghamshire)
Blindburn (Northumberland)
Blindcrake (Cumbria)
BLISS GATE (Hereford & Worcester)
BLITHBURY (Staffordshire)
Blitterlees (Cumbria)
Blockley (Gloucestershire)
Blore (Staffordshire)
Blounts Green (Staffordshire)
Blowick (Merseyside)
BLOXWICH (West Midlands)
BLUBBERHOUSES (North Yorkshire)
Blue Anchor (Somerset)
Blue Bell Hill (Kent)
Blundellsands (Merseyside)
Bluntington (Hereford & Worcester)
Blunts Green (Warwickshire)
BLURTON (Staffordshire)
Blyborough (Lincolnshire)
Blymhill (Staffordshire)
Blymhill Lawn (Staffordshire)
BLYTH (Northumberland)
BLYTH (Nottinghamshire)
BLYTHE BRIDGE (Staffordshire)
BLYTHE END (Warwickshire)
Blythe Marsh (Staffordshire)
Blyton (Lincolnshire)
BOAR'S HEAD (Greater Manchester)
Boarley (Kent)
BOARSGREAVE (Lancashire)
Bobbing (Kent)
Bobbington (Staffordshire)
Bockleton (Hereford & Worcester)
Boddington (Gloucestershire)
Bodedern (Anglesey)

Bodelwyddan (Denbighshire)
Bodenham (Hereford & Worcester)
Bodenham Moor (Hereford & Worcester)
Bodewryd (Anglesey)
Bodfari (Denbighshire)
Bodffordd (Anglesey)
Bodorgan (Anglesey)
Bodsham Green (Kent)
BODYMOOR HEATH (Warwickshire)
Bolam (Durham)
BOLAM (Northumberland)
BOLD HEATH (Merseyside)
Boldmere (West Midlands)
BOLDON (Tyne & Wear)
BOLDON COLLIERY (Tyne & Wear)
Boldron (Durham)
Bole (Nottinghamshire)
BOLE HILL (Derbyshire)
Bolehill (Derbyshire)
BOLLINGTON (Cheshire)
BOLLINGTON CROSS (Cheshire)
Bollow (Gloucestershire)
BOLSOVER (Derbyshire)
BOLSTER MOOR (West Yorkshire)
BOLSTERSTONE (South Yorkshire)
Boltby (North Yorkshire)
Bolton (Cumbria)
Bolton (East Riding of Yorkshire)
BOLTON (Greater Manchester)
Bolton (Northumberland)
BOLTON ABBEY (North Yorkshire)
BOLTON BRIDGE (North Yorkshire)
Bolton by Bowland (Lancashire)
BOLTON HALL (North Yorkshire)
Bolton le Sands (Lancashire)
BOLTON LOW HOUSES (Cumbria)
BOLTON NEW HOUSES (Cumbria)
Bolton Percy (North Yorkshire)
Bolton Town End (Lancashire)
BOLTON UPON DEARNE (South Yorkshire)
BOLTON-ON-SWALE (North Yorkshire)
BOLTONFELLEND (Cumbria)
BOLTONGATE (Cumbria)
Bomere Heath (Shropshire)
Bonby (Lincolnshire)
Boncath (Pembrokeshire)
Bond's Green (Hereford & Worcester)
Bonds (Lancashire)
BONEHILL (Staffordshire)
BONEY HAY (Staffordshire)
Boningale (Shropshire)
Bonnington (Kent)
Bonsall (Derbyshire)
Bont (Monmouthshire)
Bont-Dolgadfan (Powys)
Bonthorpe (Lincolnshire)
Bontuchel (Denbighshire)
Bonvilston (Vale of Glamorgan)
Bonwm (Denbighshire)
BONYMAEN (Swansea)
Booley (Shropshire)
BOON HILL (Staffordshire)
Boosbeck (North Yorkshire)
Boot (Cumbria)
Booth (East Riding of Yorkshire)
BOOTH (West Yorkshire)
BOOTH GREEN (Cheshire)
BOOTH TOWN (West Yorkshire)
Boothby Graffoe (Lincolnshire)
Boothby Pagnell (Lincolnshire)
BOOTHSTOWN (Greater Manchester)
Bootle (Cumbria)
Bootle (Merseyside)
Boots Green (Cheshire)
BOOZE (North Yorkshire)
Boraston (Shropshire)
Borden (Kent)
Border (Cumbria)
BORDLEY (North Yorkshire)
Boreton (Cheshire)
Borough Green (Kent)
Boroughbridge (North Yorkshire)
BORRAS HEAD (Wrexham)
Borrowash (Derbyshire)
BORROWBY (North Yorkshire)
Borrowdale (Cumbria)
Borstal (Kent)
Borwick (Lancashire)
Borwick Lodge (Cumbria)
Borwick Rails (Cumbria)
Bosbury (Hereford & Worcester)

Bosherston (Pembrokeshire)
Bosley (Cheshire)
Bossall (North Yorkshire)
Bossingham (Kent)
Bossington (Somerset)
Bostock Green (Cheshire)
Boston (Lincolnshire)
Boston Spa (West Yorkshire)
Botcheston (Leicestershire)
BOTHAL (Northumberland)
BOTHAMSALL (Nottinghamshire)
BOTHEL (Cumbria)
Botolph's Bridge (Kent)
Bottesford (Leicestershire)
Bottesford (Lincolnshire)
BOTTOM O' TH' MOOR (Greater Manchester)
Bottom of Hutton (Lancashire)
Bottoms (West Yorkshire)
BOTTS GREEN (Warwickshire)
Bough Beech (Kent)
Boughrood (Powys)
Boughspring (Gloucestershire)
BOUGHTON (Nottinghamshire)
Boughton Aluph (Kent)
Boughton Green (Kent)
Boughton Lees (Kent)
Boughton Malherbe (Kent)
Boughton Monchelsea (Kent)
Boughton Street (Kent)
BOULBY (North Yorkshire)
Boulder Clough (West Yorkshire)
Bouldon (Shropshire)
Boulmer (Northumberland)
BOULSTON (Pembrokeshire)
Boultham (Lincolnshire)
Bourne (Lincolnshire)
Bournebrook (West Midlands)
Bournes Green (Gloucestershire)
Bournheath (Hereford & Worcester)
BOURNMOOR (Durham)
Bournstream (Gloucestershire)
Bournville (West Midlands)
Bourton (Shropshire)
Bourton (Somerset)
BOURTON ON DUNSMORE (Warwickshire)
Bourton-on-the- Hill (Gloucestershire)
Bourton-on-the-Water (Gloucestershire)
Boustead Hill (Cumbria)
Bouth (Cumbria)
BOUTHWAITE (North Yorkshire)
Bouts (Hereford & Worcester)
Bow (Cumbria)
Bowbank (Durham)
Bowbridge (Gloucestershire)
BOWBURN (Durham)
Bowdon (Greater Manchester)
BOWER ASHTON (Bristol)
Bower Hinton (Somerset)
BOWER'S ROW (West Yorkshire)
BOWERS (Staffordshire)
Bowes (Durham)
Bowgreave (Lancashire)
Bowker's Green (Lancashire)
Bowland Bridge (Cumbria)
BOWLEE (Greater Manchester)
Bowley (Hereford & Worcester)
Bowley Town (Hereford & Worcester)
BOWLING (West Yorkshire)
BOWLING BANK (Wrexham)
Bowling Green (Hereford & Worcester)
Bowmanstead (Cumbria)
Bowness-on-Solway (Cumbria)
Bowness-on-Windermere (Cumbria)
Bowscale (Cumbria)
BOWSDEN (Northumberland)
Box (Gloucestershire)
Boxbush (Gloucestershire)
Boxholm (Lincolnshire)
Boxley (Kent)
Boxwell (Gloucestershire)
Boyden Gate (Kent)
Boylestone (Derbyshire)
Boynton (East Riding of Yorkshire)
BOYTHORPE (Derbyshire)
Brabourne (Kent)
Brabourne Lees (Kent)
Braceborough (Lincolnshire)
Bracebridge Heath (Lincolnshire)
Bracebridge Low Fields (Lincolnshire)
Braceby (Lincolnshire)
Bracewell (Lancashire)

See paras. 2 and 4 of the User Guide 2003 if you can't find your place name.

BRACKENFIELD (Derbyshire)
BRACKENTHWAITE (Cumbria)
Brackenthwaite (North Yorkshire)
Bracon (Lincolnshire)
Bradbourne (Derbyshire)
BRADBURY (Durham)
BRADELEY (Staffordshire)
BRADFIELD (South Yorkshire)
Bradfield Green (Cheshire)
Bradford (Northumberland)
BRADFORD (West Yorkshire)
Bradford-on-Tone (Somerset)
Bradley (Cheshire)
Bradley (Derbyshire)
Bradley (Hereford & Worcester)
Bradley (Lincolnshire)
BRADLEY (North Yorkshire)
Bradley (Staffordshire)
BRADLEY (West Midlands)
BRADLEY (West Yorkshire)
BRADLEY (Wrexham)
Bradley Green (Cheshire)
Bradley Green (Hereford & Worcester)
Bradley Green (Somerset)
BRADLEY GREEN (Warwickshire)
Bradley in the Moors (Staffordshire)
Bradley Stoke (Gloucestershire)
Bradmore (Nottinghamshire)
Bradney (Somerset)
Bradnop (Staffordshire)
Bradnor Green (Hereford & Worcester)
BRADSHAW (Greater Manchester)
BRADSHAW (West Yorkshire)
Bradwall Green (Cheshire)
Bradwell (Derbyshire)
Brafferton (Durham)
Brafferton (North Yorkshire)
Brailsford (Derbyshire)
Brailsford Green (Derbyshire)
Brain's Green (Gloucestershire)
Braithwaite (Cumbria)
BRAITHWAITE (West Yorkshire)
BRAITHWELL (South Yorkshire)
BRAKEN HILL (West Yorkshire)
BRAMCOTE (Nottinghamshire)
Bramcote (Warwickshire)
Bramhall (Greater Manchester)
Bramham (West Yorkshire)
BRAMHOPE (West Yorkshire)
BRAMLEY (Derbyshire)
BRAMLEY (South Yorkshire)
BRAMLEY (West Yorkshire)
BRAMLEY HEAD (North Yorkshire)
BRAMLING (Kent)
BRAMPTON (Cumbria)
Brampton (Lincolnshire)
BRAMPTON (South Yorkshire)
Brampton Abbotts (Hereford & Worcester)
Brampton Bryan (Hereford & Worcester)
BRAMPTON-EN-LE-MORTHEN (South Yorkshire)
Bramshall (Staffordshire)
Bramwell (Somerset)
BRANCEPETH (Durham)
BRANCH END (Northumberland)
Brand End (Lincolnshire)
Brand Green (Gloucestershire)
Brandesburton (East Riding of Yorkshire)
BRANDON (Durham)
Brandon (Lincolnshire)
Brandon (Northumberland)
Brandon (Warwickshire)
Brandsby (North Yorkshire)
Brandy Wharf (Lincolnshire)
Bransby (Lincolnshire)
Bransford (Hereford & Worcester)
Bransley (Shropshire)
Branson's Cross (Hereford & Worcester)
Branston (Leicestershire)
Branston (Lincolnshire)
BRANSTON (Staffordshire)
Branston Booths (Lincolnshire)
Brant Broughton (Lincolnshire)
BRANTHWAITE (Cumbria)
Brantingham (East Riding of Yorkshire)
Branton (Northumberland)
BRANTON (South Yorkshire)
Branton Green (North Yorkshire)
Branxton (Northumberland)
Brassey Green (Cheshire)
Brassington (Derbyshire)
Brasted (Kent)

Brasted Chart (Kent)
Bratoft (Lincolnshire)
Brattleby (Lincolnshire)
BRATTON (Shropshire)
Bratton (Somerset)
Bratton Seymour (Somerset)
Braunston (Lincolnshire)
Braunstone (Leicestershire)
Brawby (North Yorkshire)
Brawdy (Pembrokeshire)
Braworth (North Yorkshire)
Braystones (Cumbria)
BRAYTHORN (North Yorkshire)
BRAYTON (North Yorkshire)
Breach (Kent)
Breaden Heath (Shropshire)
Breadsall (Derbyshire)
Breadstone (Gloucestershire)
Breadward (Hereford & Worcester)
BREAM (Gloucestershire)
Brean (Somerset)
BREARLEY (West Yorkshire)
Brearton (North Yorkshire)
Breaston (Derbyshire)
Brechfa (Carmarthenshire)
Breckfield (Merseyside)
Brecon (Powys)
BREDBURY (Greater Manchester)
Bredenbury (Hereford & Worcester)
Bredgar (Kent)
Bredhurst (Kent)
Bredon (Hereford & Worcester)
Bredon's Hardwick (Hereford & Worcester)
Bredon's Norton (Hereford & Worcester)
Bredwardine (Hereford & Worcester)
BREEDON ON THE HILL (Leicestershire)
BREIGHTMET (Greater Manchester)
BREIGHTON (East Riding of Yorkshire)
Breinton (Hereford & Worcester)
Brenchley (Kent)
Brendon Hill (Somerset)
BRENKLEY (Tyne & Wear)
Brent Knoll (Somerset)
Brentingby (Leicestershire)
Brentry (Bristol)
Brenzett (Kent)
Brenzett Green (Kent)
BRERETON (Staffordshire)
Brereton Green (Cheshire)
Brereton Heath (Cheshire)
BRERETON HILL (Staffordshire)
BRETBY (Derbyshire)
Bretford (Warwickshire)
Bretforton (Hereford & Worcester)
Bretherdale Head (Cumbria)
Bretherton (Lancashire)
Bretton (Derbyshire)
BRETTON (Flintshire)
Brewood (Staffordshire)
BRICK HOUSES (South Yorkshire)
Bricklehampton (Hereford & Worcester)
Bridekirk (Cumbria)
Bridell (Pembrokeshire)
Bridge (Kent)
Bridge End (Cumbria)
BRIDGE END (Durham)
Bridge End (Lincolnshire)
BRIDGE END (Northumberland)
Bridge Fields (Leicestershire)
Bridge Hewick (North Yorkshire)
Bridge Sollers (Hereford & Worcester)
Bridge Trafford (Cheshire)
BRIDGEFOOT (Cumbria)
Bridgehampton (Somerset)
BRIDGEHILL (Durham)
BRIDGEHOUSE GATE (North Yorkshire)
Bridgend (Bridgend)
Bridgend (Cumbria)
Bridges (Shropshire)
Bridgetown (Somerset)
BRIDGEYATE (Gloucestershire)
BRIDGNORTH (Shropshire)
BRIDGTOWN (Staffordshire)
Bridgwater (Somerset)
Bridlington (East Riding of Yorkshire)
Bridstow (Hereford & Worcester)
BRIERFIELD (Lancashire)
BRIERLEY (Gloucestershire)
Brierley (Hereford & Worcester)
BRIERLEY (West Yorkshire)
BRIERLEY HILL (West Midlands)

Brierton (Durham)
Briery (Cumbria)
Brigg (Lincolnshire)
BRIGGSWATH (North Yorkshire)
BRIGHAM (Cumbria)
Brigham (East Riding of Yorkshire)
BRIGHOUSE (West Yorkshire)
Brightgate (Derbyshire)
BRIGHTHOLMLEE (Derbyshire)
Brighton le Sands (Merseyside)
Brignall (Durham)
Brigsley (Lincolnshire)
Brigsteer (Cumbria)
Brilley (Hereford & Worcester)
Brimfield (Hereford & Worcester)
Brimfield Cross (Hereford & Worcester)
BRIMINGTON (Derbyshire)
Brimpsfield (Gloucestershire)
Brimscombe (Gloucestershire)
Brimstage (Merseyside)
BRINCLIFFE (South Yorkshire)
Brind (East Riding of Yorkshire)
Brindham (Somerset)
BRINDLE (Lancashire)
Brineton (Staffordshire)
Bringhurst (Leicestershire)
Brinkely (Nottinghamshire)
Brinkhill (Lincolnshire)
Brinklow (Warwickshire)
BRINSCALL (Lancashire)
Brinscombe (Somerset)
Brinsea (Somerset)
BRINSLEY (Nottinghamshire)
Brinsop (Hereford & Worcester)
BRINSWORTH (South Yorkshire)
Brisco (Cumbria)
BRISLINGTON (Bristol)
Brissenden Green (Kent)
BRISTOL (Bristol) - check district
BRITANNIA (Lancashire)
BRITHDIR (Caerphilly)
BRITISH (Torfaen)
British Legion Village (Kent)
BRITON FERRY (Neath Port Talbot)
Broad Alley (Hereford & Worcester)
Broad Campden (Gloucestershire)
BROAD CARR (West Yorkshire)
BROAD CLOUGH (Lancashire)
Broad Ford (Kent)
Broad Green (Hereford & Worcester)
Broad Green (Merseyside)
BROAD HAVEN (Pembrokeshire)
Broad Marston (Hereford & Worcester)
BROAD MEADOW (Staffordshire)
Broad Oak (Cumbria)
Broad Oak (Hereford & Worcester)
BROAD OAK (Merseyside)
Broad Street (Kent)
BROADBOTTOM (Greater Manchester)
BROADFIELD (Pembrokeshire)
Broadgate (Lincolnshire)
Broadheath (Greater Manchester)
Broadheath (Hereford & Worcester)
Broadholme (Nottinghamshire)
Broadlay (Carmarthenshire)
BROADLEY (Greater Manchester)
BROADMOOR (Gloucestershire)
BROADMOOR (Pembrokeshire)
Broadoak (Gloucestershire)
BROADOAK (Wrexham)
Broadstairs (Kent)
Broadstone (Monmouthshire)
Broadstone (Shropshire)
Broadwas (Hereford & Worcester)
Broadwaters (Hereford & Worcester)
Broadway (Carmarthenshire)
Broadway (Hereford & Worcester)
Broadway (Pembrokeshire)
Broadway (Somerset)
BROADWELL (Gloucestershire)
Broadwell (Warwickshire)
Brobury (Hereford & Worcester)
Brock (Lancashire)
Brockamin (Hereford & Worcester)
Brockencote (Hereford & Worcester)
Brockhampton (Gloucestershire)
Brockhampton (Hereford & Worcester)
BROCKHOLES (West Yorkshire)
Brockhurst (Derbyshire)
Brockhurst (Warwickshire)
BROCKLEBANK (Cumbria)

Brocklesby (Lincolnshire)
Brockley (Somerset)
Brockleymoor (Cumbria)
BROCKMOOR (West Midlands)
BROCKTON (Shropshire)
Brockton (Staffordshire)
Brockweir (Gloucestershire)
Brockworth (Gloucestershire)
BROCTON (Staffordshire)
BRODSWORTH (South Yorkshire)
BROKEN CROSS (Cheshire)
Bromborough (Merseyside)
BROMFIELD (Cumbria)
Bromfield (Shropshire)
Bromley (Shropshire)
BROMLEY (South Yorkshire)
BROMLEY (West Midlands)
Bromlow (Shropshire)
Brompton (Kent)
Brompton (North Yorkshire)
Brompton (Shropshire)
Brompton Ralph (Somerset)
Brompton Regis (Somerset)
Brompton-on-Swale (North Yorkshire)
Bromsash (Hereford & Worcester)
Bromsberrow (Gloucestershire)
Bromsberrow Heath (Gloucestershire)
Bromsgrove (Hereford & Worcester)
Bromstead Heath (Staffordshire)
Bromyard (Hereford & Worcester)
Bromyard Downs (Hereford & Worcester)
Broncroft (Shropshire)
BRONINGTON (Wrexham)
Bronllys (Powys)
Bronwydd (Carmarthenshire)
Bronydd (Powys)
Bronygarth (Shropshire)
Brook (Carmarthenshire)
Brook (Kent)
Brook House (Denbighshire)
Brook Street (Kent)
Brookhampton (Somerset)
BROOKHOUSE (Lancashire)
BROOKHOUSE (South Yorkshire)
Brookhouse Green (Cheshire)
BROOKHOUSES (Derbyshire)
Brookland (Kent)
Brooklands (Greater Manchester)
Brooks (Powys)
Brooks End (Kent)
Brooksby (Leicestershire)
Brookthorpe (Gloucestershire)
BROOM (South Yorkshire)
Broom (Warwickshire)
Broom Hill (Hereford & Worcester)
BROOM HILL (Nottinghamshire)
BROOM HILL (South Yorkshire)
Broom Street (Kent)
Broom's Green (Gloucestershire)
Broome (Hereford & Worcester)
Broome (Shropshire)
Broome Park (Northumberland)
Broomedge (Cheshire)
Broomfield (Kent)
Broomfield (Somerset)
Broomfields (Shropshire)
Broomfleet (East Riding of Yorkshire)
BROOMHAUGH (Northumberland)
Broomhill (Northumberland)
Broomhill Green (Cheshire)
BROOMLEY (Northumberland)
BROSELEY (Shropshire)
Brotherhouse Bar (Lincolnshire)
BROTHERLEE (Durham)
Brothertoft (Lincolnshire)
BROTHERTON (North Yorkshire)
Brotton (North Yorkshire)
BROUGH (Cumbria)
Brough (Derbyshire)
Brough (East Riding of Yorkshire)
Brough (Nottinghamshire)
BROUGH SOWERBY (Cumbria)
Broughall (Shropshire)
BROUGHTON (Flintshire)
BROUGHTON (Greater Manchester)
Broughton (Lancashire)
Broughton (North Yorkshire)
Broughton (Staffordshire)
Broughton Astley (Leicestershire)
Broughton Beck (Cumbria)
Broughton Green (Hereford & Worcester)

Broughton Hackett (Hereford & Worcester)
Broughton Mills (Cumbria)
BROUGHTON MOOR (Cumbria)
Broughton Tower (Cumbria)
Broughton-in-Furness (Cumbria)
Brow End (Cumbria)
Brown Edge (Lancashire)
BROWN EDGE (Staffordshire)
Brown Heath (Cheshire)
BROWN LEES (Staffordshire)
Brown's Green (West Midlands)
BROWNBER (Cumbria)
Brownheath (Shropshire)
BROWNHILLS (West Midlands)
BROWNIESIDE (Northumberland)
BROWNLOW HEATH (Cheshire)
BROWNRIGG (Cumbria)
Brownsover (Warwickshire)
Broxa (North Yorkshire)
BROXFIELD (Northumberland)
Broxton (Cheshire)
Broxwood (Hereford & Worcester)
Bruera (Cheshire)
Brumby (Lincolnshire)
Brund (Staffordshire)
Brunslow (Shropshire)
BRUNTCLIFFE (West Yorkshire)
BRUNTHWAITE (West Yorkshire)
Bruntingthorpe (Leicestershire)
BRUNTON (Northumberland)
Brushford (Somerset)
Bruton (Somerset)
Bryan's Green (Hereford & Worcester)
BRYMBO (Wrexham)
Brympton (Somerset)
Bryn (Cheshire)
BRYN (Greater Manchester)
BRYN (Neath Port Talbot)
Bryn (Shropshire)
Bryn Du (Anglesey)
BRYN GATES (Lancashire)
BRYN GOLAU (Rhondda Cynon Taff)
Bryn Saith Marchog (Denbighshire)
BRYN-COCH (Neath Port Talbot)
Bryn-henllan (Pembrokeshire)
Bryn-newydd (Denbighshire)
Bryn-penarth (Powys)
BRYN-Y-BAAL (Flintshire)
BRYN-YR-EOS (Wrexham)
BRYNAMAN (Carmarthenshire)
Brynberian (Pembrokeshire)
BRYNBRYDDAN (Neath Port Talbot)
BRYNCAE (Rhondda Cynon Taff)
BRYNCETHIN (Bridgend)
Bryneglwys (Denbighshire)
BRYNFIELDS (Wrexham)
Brynford (Flintshire)
Bryngwran (Anglesey)
Bryngwyn (Monmouthshire)
Bryngwyn (Powys)
Bryning (Lancashire)
BRYNITHEL (Blaenau Gwent)
BRYNMAWR (Blaenau Gwent)
BRYNMENYN (Bridgend)
BRYNMILL (Swansea)
BRYNNA (Rhondda Cynon Taff)
Brynrefail (Anglesey)
BRYNSADLER (Rhondda Cynon Taff)
Brynsiencyn (Anglesey)
Brynteg (Anglesey)
BUARTH-DRAW (Flintshire)
Bubbenhall (Warwickshire)
BUBWITH (East Riding of Yorkshire)
Buckabank (Cumbria)
BUCKDEN (North Yorkshire)
Buckholt (Monmouthshire)
Buckland (Gloucestershire)
BUCKLAND (Kent)
BUCKLAND DINHAM (Somerset)
Buckland St. Mary (Somerset)
BUCKLEY (Flintshire)
Buckley Green (Warwickshire)
BUCKLEY MOUNTAIN (Flintshire)
Bucklow Hill (Cheshire)
Buckminster (Leicestershire)
Bucknall (Lincolnshire)
BUCKNALL (Staffordshire)
Bucknell (Shropshire)
Buckton (East Riding of Yorkshire)
Buckton (Hereford & Worcester)
BUCKTON (Northumberland)

Budbrooke (Warwickshire)
BUDBY (Nottinghamshire)
Buddileigh (Staffordshire)
BUDLE (Northumberland)
Buerton (Cheshire)
Buglawton (Cheshire)
Bugthorpe (East Riding of Yorkshire)
BUILDWAS (Shropshire)
Builth Road (Powys)
Builth Wells (Powys)
Bulby (Lincolnshire)
Bulkeley (Cheshire)
BULKINGTON (Warwickshire)
Bull Bay (Anglesey)
Bullamore (North Yorkshire)
BULLBRIDGE (Derbyshire)
Bulley (Gloucestershire)
BULLGILL (Cumbria)
Bullinghope (Hereford & Worcester)
Bullington (Lincolnshire)
Bullockstone (Kent)
Bulmer (North Yorkshire)
Bulterley Heath (Staffordshire)
BULWELL (Nottinghamshire)
Bunbury (Cheshire)
Bunbury Heath (Cheshire)
Bunker's Hill (Lincolnshire)
Bunny (Nottinghamshire)
Bupton (Derbyshire)
Burbage (Derbyshire)
Burbage (Leicestershire)
Burcher (Hereford & Worcester)
Burcot (Hereford & Worcester)
Burcote (Shropshire)
Burdale (North Yorkshire)
Burford (Hereford & Worcester)
Burgh by Sands (Cumbria)
Burgh Le Marsh (Lincolnshire)
Burgh on Bain (Lincolnshire)
Burghill (Hereford & Worcester)
BURGHWALLIS (South Yorkshire)
Burham (Kent)
Burland (Cheshire)
Burleigh (Gloucestershire)
Burley (Shropshire)
Burley Gate (Hereford & Worcester)
BURLEY IN WHARFEDALE (West Yorkshire)
BURLEY WOOD HEAD (West Yorkshire)
Burleydam (Cheshire)
Burlingjobb (Powys)
BURLINGTON (Shropshire)
Burlton (Shropshire)
Burmarsh (Kent)
Burmington (Warwickshire)
BURN (North Yorkshire)
BURN CROSS (South Yorkshire)
Burn Naze (Lancashire)
Burnage (Greater Manchester)
Burnaston (Derbyshire)
Burnbanks (Cumbria)
Burnby (East Riding of Yorkshire)
BURNDEN (Greater Manchester)
BURNEDGE (Greater Manchester)
Burneside (Cumbria)
Burneston (North Yorkshire)
Burnett (Somerset)
Burnham (Lincolnshire)
Burnham-on-Sea (Somerset)
Burnhill Green (Staffordshire)
BURNHOPE (Durham)
Burniston (North Yorkshire)
BURNLEY (Lancashire)
BURNOPFIELD (Durham)
Burnrigg (Cumbria)
BURNSALL (North Yorkshire)
BURNT HOUSES (Durham)
BURNT YATES (North Yorkshire)
Burntheath (Derbyshire)
BURNTWOOD (Staffordshire)
BURNTWOOD GREEN (Staffordshire)
Burnworthy (Somerset)
Burradon (Northumberland)
BURRADON (Tyne & Wear)
Burrells (Cumbria)
Burrill (North Yorkshire)
Burringham (Lincolnshire)
Burrington (Hereford & Worcester)
Burrington (Somerset)
Burrough on the Hill (Leicestershire)
Burrow (Lancashire)
Burrow (Somerset)

See paras. 2 and 4 of the User Guide 2003 if you can't find your place name.

Burrow Bridge (Somerset)
Burry (Swansea)
BURRY PORT (Carmarthenshire)
Burrygreen (Swansea)
BURSCOUGH (Lancashire)
BURSCOUGH BRIDGE (Lancashire)
Bursea (East Riding of Yorkshire)
Burshill (East Riding of Yorkshire)
BURSLEM (Staffordshire)
BURSTON (Staffordshire)
Burstwick (East Riding of Yorkshire)
BURTERSETT (North Yorkshire)
Burtholme (Cumbria)
Burthwaite (Cumbria)
Burtle Hill (Somerset)
Burtoft (Lincolnshire)
Burton (Cheshire)
Burton (Lincolnshire)
BURTON (Northumberland)
Burton (Pembrokeshire)
Burton (Somerset)
Burton Agnes (East Riding of Yorkshire)
Burton Coggles (Lincolnshire)
Burton Dassett (Warwickshire)
Burton Fleming (East Riding of Yorkshire)
BURTON GREEN (Warwickshire)
BURTON GREEN (Wrexham)
Burton Hastings (Warwickshire)
BURTON IN LONSDALE (North Yorkshire)
BURTON JOYCE (Nottinghamshire)
Burton Lazars (Leicestershire)
Burton Leonard (North Yorkshire)
Burton on the Wolds (Leicestershire)
Burton Overy (Leicestershire)
Burton Pedwardine (Lincolnshire)
Burton Pidsea (East Riding of Yorkshire)
BURTON SALMON (North Yorkshire)
Burton upon Stather (Lincolnshire)
BURTON UPON TRENT (Staffordshire)
Burton-in-Kendal (Cumbria)
BURTONWOOD (Cheshire)
Burwardsley (Cheshire)
BURWARTON (Shropshire)
Burwell (Lincolnshire)
Burwen (Anglesey)
BURY (Greater Manchester)
Bury (Somerset)
Burythorpe (North Yorkshire)
Bush Bank (Hereford & Worcester)
BUSHBURY (West Midlands)
Bushby (Leicestershire)
Bushley (Hereford & Worcester)
Bushley Green (Hereford & Worcester)
BUSK (Cumbria)
Buslingthorpe (Lincolnshire)
Bussage (Gloucestershire)
Bussex (Somerset)
Butcher Hill (West Yorkshire)
Butcombe (Somerset)
Butleigh (Somerset)
Butleigh Wootton (Somerset)
BUTLER'S HILL (Nottinghamshire)
BUTLERS GREEN (Staffordshire)
Butlers Marston (Warwickshire)
Butt Green (Cheshire)
BUTT LANE (Staffordshire)
Buttercrambe (North Yorkshire)
BUTTERKNOWLE (Durham)
BUTTERLEY (Derbyshire)
Buttermere (Cumbria)
BUTTERSHAW (West Yorkshire)
BUTTERTON (Staffordshire)
BUTTERWICK (Durham)
Butterwick (Lincolnshire)
Butterwick (North Yorkshire)
Buttington (Powys)
BUTTONBRIDGE (Shropshire)
BUTTONOAK (Shropshire)
BUXTON (Derbyshire)
Bwlch (Powys)
Bwlch-y-cibau (Powys)
Bwlch-y-ffridd (Powys)
Bwlch-y-groes (Pembrokeshire)
Bwlch-y-sarnau (Powys)
Bwlchgwyn (Wrexham)
Bwlchnewydd (Carmarthenshire)
Bwlchyddar (Powys)
BWLCHYMYRDD (Swansea)
BYERMOOR (Tyne & Wear)
BYERS GARTH (Durham)
BYERS GREEN (Durham)

Byford (Hereford & Worcester)
BYKER (Tyne & Wear)
Byley (Cheshire)
BYNEA (Carmarthenshire)
Byrness (Northumberland)
Byton (Hereford & Worcester)
Bywell (Northumberland)

C

Cabourne (Lincolnshire)
Cabus (Lancashire)
Cadeby (Leicestershire)
CADEBY (South Yorkshire)
CADISHEAD (Greater Manchester)
CADLE (Swansea)
Cadley (Lancashire)
Cadney (Lincolnshire)
Cadole (Flintshire)
Cadoxton (Vale of Glamorgan)
CADOXTON JUXTA-NEATH (Neath Port Talbot)
Cadwst (Denbighshire)
CAE'R BRYN (Carmarthenshire)
CAE'R-BONT (Powys)
CAEHOPKIN (Powys)
Caenby (Lincolnshire)
Caenby Corner (Lincolnshire)
Caeo (Carmarthenshire)
Caer Farchell (Pembrokeshire)
CAERAU (Bridgend)
Caerau (Cardiff)
Caergeiliog (Anglesey)
Caergwrle (Flintshire)
Caerleon (Newport)
CAERPHILLY (Caerphilly)
Caersws (Powys)
Caerwent (Monmouthshire)
Caerwys (Flintshire)
Caggle Street (Monmouthshire)
Caim (Anglesey)
Caistor (Lincolnshire)
Cakebole (Hereford & Worcester)
Calceby (Lincolnshire)
Calcot (Flintshire)
Calcot (Gloucestershire)
Calcott (Kent)
Calcott (Shropshire)
Calcutt (North Yorkshire)
CALDBECK (Cumbria)
CALDBERGH (North Yorkshire)
Calder Bridge (Cumbria)
CALDER GROVE (West Yorkshire)
Calder Vale (Lancashire)
CALDERBROOK (Greater Manchester)
CALDERMORE (Greater Manchester)
Caldicot (Monmouthshire)
CALDWELL (North Yorkshire)
Caldy (Merseyside)
Caledfwlch (Carmarthenshire)
California (Derbyshire)
CALKE (Derbyshire)
Callaly (Northumberland)
Callaughton (Shropshire)
Callingwood (Staffordshire)
Callow (Hereford & Worcester)
Callow End (Hereford & Worcester)
CALLOW HILL (Hereford & Worcester)
Callows Grave (Hereford & Worcester)
Calmsden (Gloucestershire)
CALOW (Derbyshire)
Calthwaite (Cumbria)
Calton (North Yorkshire)
Calton (Staffordshire)
Calton Green (Staffordshire)
Calveley (Cheshire)
Calver (Derbyshire)
Calver Hill (Hereford & Worcester)
Calver Sough (Derbyshire)
Calverhall (Shropshire)
CALVERLEY (West Yorkshire)
CALVERTON (Nottinghamshire)
Calvo (Cumbria)
Cam (Gloucestershire)
CAMBLESFORTH (North Yorkshire)
CAMBO (Northumberland)
CAMBOIS (Northumberland)
Cambridge (Gloucestershire)
Cameley (Somerset)
Camer's Green (Hereford & Worcester)
CAMERTON (Cumbria)
CAMERTON (Somerset)

Cammeringham (Lincolnshire)
Camp The (Gloucestershire)
CAMPERDOWN (Tyne & Wear)
CAMPSALL (South Yorkshire)
Camrose (Pembrokeshire)
Canal Foot (Cumbria)
Canaston Bridge (Pembrokeshire)
Candlesby (Lincolnshire)
CANDOVER GREEN (Shropshire)
CANKLOW (South Yorkshire)
CANLEY (West Midlands)
Cannington (Somerset)
CANNOCK (Staffordshire)
CANNOCK WOOD (Staffordshire)
Cannon Bridge (Hereford & Worcester)
Canon Frome (Hereford & Worcester)
Canon Pyon (Hereford & Worcester)
Canterbury (Kent)
CANTLEY (South Yorkshire)
Cantlop (Shropshire)
Canton (Cardiff)
CANTSFIELD (Lancashire)
Canwick (Lincolnshire)
Capel (Kent)
Capel Coch (Anglesey)
Capel Dewi (Carmarthenshire)
Capel Gwyn (Anglesey)
Capel Gwyn (Carmarthenshire)
Capel Gwynfe (Carmarthenshire)
CAPEL HENDRE (Carmarthenshire)
CAPEL ISAAC (Carmarthenshire)
Capel Iwan (Carmarthenshire)
Capel le Ferne (Kent)
Capel Llanilltern (Cardiff)
Capel Mawr (Anglesey)
Capel Parc (Anglesey)
Capel-y-ffin (Powys)
Capenhurst (Cheshire)
Capernwray (Lancashire)
CAPHEATON (Northumberland)
Capstone (Kent)
Capton (Somerset)
Car Colston (Nottinghamshire)
CARBROOK (South Yorkshire)
CARBURTON (Nottinghamshire)
CARCROFT (South Yorkshire)
Cardeston (Shropshire)
Cardewlees (Cumbria)
Cardiff (Cardiff)
CARDINGTON (Shropshire)
Cardurnock (Cumbria)
Careby (Lincolnshire)
Carew (Pembrokeshire)
Carew Cheriton (Pembrokeshire)
CAREW NEWTON (Pembrokeshire)
Carey (Hereford & Worcester)
Cargo (Cumbria)
Carham (Northumberland)
Carhampton (Somerset)
Cark (Cumbria)
Carlbury (Durham)
Carlby (Lincolnshire)
Carlcroft (Northumberland)
CARLECOTES (South Yorkshire)
CARLESMOOR (North Yorkshire)
Carleton (Cumbria)
Carleton (Lancashire)
CARLETON (North Yorkshire)
CARLETON (West Yorkshire)
Carlin How (North Yorkshire)
CARLINGCOTT (Somerset)
Carlisle (Cumbria)
Carlton (Cumbria)
Carlton (Durham)
Carlton (Leicestershire)
CARLTON (North Yorkshire)
CARLTON (Nottinghamshire)
CARLTON (South Yorkshire)
CARLTON (West Yorkshire)
Carlton Curlieu (Leicestershire)
CARLTON HUSTHWAITE (North Yorkshire)
CARLTON IN LINDRICK (Nottinghamshire)
Carlton Miniott (North Yorkshire)
Carlton Scroop (Lincolnshire)
Carlton-le-Moorland (Lincolnshire)
CARLTON-ON-TRENT (Nottinghamshire)
Carmarthen (Carmarthenshire)
Carmel (Anglesey)
CARMEL (Carmarthenshire)
Carmel (Flintshire)
Carnaby (East Riding of Yorkshire)

See paras. 2 and 4 of the User Guide 2003 if you can't find your place name.

Carnforth (Lancashire)
Carnhedryn (Pembrokeshire)
Carno (Powys)
CAROL GREEN (West Midlands)
CARPERBY (North Yorkshire)
CARR (Greater Manchester)
CARR (South Yorkshire)
CARR GATE (West Yorkshire)
CARR SHIELD (Northumberland)
CARR VALE (Derbyshire)
CARRBROOK (Greater Manchester)
Carreglefn (Anglesey)
Carrhouse (Lincolnshire)
Carrington (Greater Manchester)
Carrington (Lincolnshire)
Carrog (Denbighshire)
Carrow Hill (Monmouthshire)
CARRVILLE (Durham)
CARRYCOATS HALL (Northumberland)
Carsington (Derbyshire)
Carterway Heads (Durham)
Carthorpe (North Yorkshire)
Cartington (Northumberland)
CARTLEDGE (Derbyshire)
Cartmel (Cumbria)
Cartmel Fell (Cumbria)
CARWAY (Carmarthenshire)
Carwinley (Cumbria)
Cashe's Green (Gloucestershire)
CASSOP COLLIERY (Durham)
Castell-y-bwch (Torfaen)
CASTERTON (Lancashire)
CASTLE BOLTON (North Yorkshire)
Castle Bromwich (West Midlands)
Castle Bytham (Lincolnshire)
Castle Caereinion (Powys)
CASTLE CARROCK (Cumbria)
Castle Cary (Somerset)
Castle Donington (Leicestershire)
CASTLE EDEN (Durham)
Castle Frome (Hereford & Worcester)
Castle Green (Cumbria)
CASTLE GRESLEY (Derbyshire)
Castle Hill (Kent)
Castle Morris (Pembrokeshire)
CASTLE PULVERBATCH (Shropshire)
CASTLE STREET (West Yorkshire)
Castlebythe (Pembrokeshire)
CASTLECROFT (Staffordshire)
CASTLECROFT (West Midlands)
CASTLEFORD (West Yorkshire)
Castlemartin (Pembrokeshire)
Castlemorton (Hereford & Worcester)
CASTLESIDE (Durham)
Castlethorpe (Lincolnshire)
Castleton (Derbyshire)
CASTLETON (Greater Manchester)
Castleton (Newport)
CASTLETON (North Yorkshire)
CASTLETOWN (Tyne & Wear)
Castley (North Yorkshire)
Caswell Bay (Swansea)
CAT AND FIDDLE (Cheshire)
Cat's Ash (Newport)
Catbrook (Monmouthshire)
Catch (Flintshire)
CATCHEM'S CORNER (West Midlands)
CATCHGATE (Durham)
CATCLIFFE (South Yorkshire)
Catcott (Somerset)
Catcott Burtle (Somerset)
Catforth (Lancashire)
Cathedine (Powys)
CATHERINE SLACK (West Yorkshire)
Catherine-de-Barnes (West Midlands)
CATHERTON (Shropshire)
CATLEY LANE HEAD (Greater Manchester)
Catley Southfield (Hereford & Worcester)
CATLOW (Lancashire)
CATLOWDY (Cumbria)
CATON (Lancashire)
CATON GREEN (Lancashire)
Catsgore (Somerset)
Catsham (Somerset)
Catshill (Hereford & Worcester)
Catstree (Shropshire)
Cattal (North Yorkshire)
Catterall (Lancashire)
Catteralslane (Shropshire)
Catterick (North Yorkshire)
Catterick Bridge (North Yorkshire)

Catterick Garrison (North Yorkshire)
Catterlen (Cumbria)
Catterton (North Yorkshire)
Catthorpe (Leicestershire)
Catton (Cumbria)
Catton (North Yorkshire)
Catwick (East Riding of Yorkshire)
Caudle Green (Gloucestershire)
Cauldon (Staffordshire)
Cauldon Lowe (Staffordshire)
CAULDWELL (Derbyshire)
Caunsall (Hereford & Worcester)
CAUNTON (Nottinghamshire)
Causeway End (Cumbria)
Causewayhead (Cumbria)
CAUSEY PARK (Northumberland)
CAVERSWALL (Staffordshire)
Cavil (East Riding of Yorkshire)
Cawkwell (Lincolnshire)
CAWOOD (North Yorkshire)
Cawston (Warwickshire)
Cawthorn (North Yorkshire)
CAWTHORNE (South Yorkshire)
CAYNHAM (Shropshire)
Caythorpe (Lincolnshire)
CAYTHORPE (Nottinghamshire)
Cayton (North Yorkshire)
Ceciliford (Monmouthshire)
Cefn (Newport)
CEFN BYRLE (Powys)
Cefn Canel (Powys)
Cefn Coch (Powys)
CEFN CRIBWR (Bridgend)
CEFN CROSS (Bridgend)
Cefn Mably (Caerphilly)
CEFN-BRYN-BRAIN (Carmarthenshire)
Cefn-coed-y-cymmer (Merthyr Tydfil)
Cefn-Einion (Shropshire)
CEFN-MAWR (Wrexham)
CEFN-Y-BEDD (Wrexham)
Cefn-y-pant (Carmarthenshire)
CEFNEITHIN (Carmarthenshire)
Cefngorwydd (Powys)
CEFNPENNAR (Rhondda Cynon Taff)
CEINT (Anglesey)
CELLARHEAD (Staffordshire)
Cellerton (Cumbria)
CELYNEN (Caerphilly)
Cemaes (Anglesey)
Cemmaes (Powys)
Cemmaes Road (Powys)
Cenarth (Cardiganshire)
Cerbyd (Pembrokeshire)
Cerney Wick (Gloucestershire)
Cerrigceinwen (Anglesey)
Chaceley (Gloucestershire)
Chadbury (Hereford & Worcester)
CHADDERTON (Greater Manchester)
CHADDERTON FOLD (Greater Manchester)
Chaddesden (Derbyshire)
Chaddesley Corbett (Hereford & Worcester)
Chadshunt (Warwickshire)
CHADWELL (Leicestershire)
Chadwell (Shropshire)
Chadwick (Hereford & Worcester)
Chadwick End (West Midlands)
CHADWICK GREEN (Merseyside)
Chaffcombe (Somerset)
Chainhurst (Kent)
Chalford (Gloucestershire)
Chalk (Kent)
Chalkway (Somerset)
Chalkwell (Kent)
Challock Lees (Kent)
Chambers Green (Kent)
Chandlers Cross (Hereford & Worcester)
Chantry (Somerset)
Chapel (Cumbria)
Chapel Allerton (Somerset)
CHAPEL ALLERTON (West Yorkshire)
CHAPEL CHORLTON (Staffordshire)
CHAPEL END (Warwickshire)
CHAPEL FIELD (Greater Manchester)
CHAPEL GREEN (Warwickshire)
CHAPEL HADDLESEY (North Yorkshire)
Chapel Hill (Lincolnshire)
Chapel Hill (Monmouthshire)
Chapel Hill (North Yorkshire)
Chapel Lawn (Shropshire)
CHAPEL LE DALE (North Yorkshire)
Chapel Leigh (Somerset)

Chapel Milton (Derbyshire)
Chapel St. Leonards (Lincolnshire)
Chapel Stile (Cumbria)
Chapel-en-le-Frith (Derbyshire)
Chapelgate (Lincolnshire)
Chapels (Cumbria)
CHAPELTOWN (Lancashire)
CHAPELTOWN (South Yorkshire)
Chard (Somerset)
Chard Junction (Somerset)
Chardleigh Green (Somerset)
Charfield (Gloucestershire)
Chargrove (Gloucestershire)
Charing (Kent)
Charing Heath (Kent)
Charing Hill (Kent)
Charingworth (Gloucestershire)
Charlcombe (Somerset)
Charlecote (Warwickshire)
CHARLESTOWN (Derbyshire)
CHARLESTOWN (Greater Manchester)
CHARLESTOWN (West Yorkshire)
CHARLESWORTH (Derbyshire)
Charlinch (Somerset)
Charlton (Hereford & Worcester)
CHARLTON (Northumberland)
Charlton (Shropshire)
Charlton (Somerset)
Charlton Abbots (Gloucestershire)
Charlton Adam (Somerset)
CHARLTON HILL (Shropshire)
Charlton Horethorne (Somerset)
Charlton Kings (Gloucestershire)
Charlton Mackrell (Somerset)
Charlton Musgrove (Somerset)
CHARNOCK GREEN (Lancashire)
CHARNOCK RICHARD (Lancashire)
Chart Corner (Kent)
Chart Hill (Kent)
Chart Sutton (Kent)
Charterhouse (Somerset)
Chartham (Kent)
Chartham Hatch (Kent)
Chartway Street (Kent)
CHASE TERRACE (Staffordshire)
CHASETOWN (Staffordshire)
Chatburn (Lancashire)
CHATCULL (Staffordshire)
CHATHAM (Caerphilly)
Chatham (Kent)
CHATHILL (Northumberland)
Chatley (Hereford & Worcester)
Chattenden (Kent)
CHATTERTON (Lancashire)
Chatton (Northumberland)
Chaxhill (Gloucestershire)
Cheadle (Greater Manchester)
CHEADLE (Staffordshire)
Cheadle Heath (Greater Manchester)
Cheadle Hulme (Greater Manchester)
Chebsey (Staffordshire)
Checkley (Cheshire)
CHECKLEY (Staffordshire)
Checkley Green (Cheshire)
Cheddar (Somerset)
CHEDDLETON (Staffordshire)
CHEDDLETON HEATH (Staffordshire)
Cheddon Fitzpaine (Somerset)
Chedworth (Gloucestershire)
Chedzoy (Somerset)
CHEESDEN (Greater Manchester)
Cheeseman's Green (Kent)
CHEETHAM HILL (Greater Manchester)
CHEETWOOD (Greater Manchester)
Chelford (Cheshire)
Chellaston (Derbyshire)
CHELMARSH (Shropshire)
Chelmick (Shropshire)
Chelmorton (Derbyshire)
Chelmsley Wood (West Midlands)
Chelston (Somerset)
Cheltenham (Gloucestershire)
CHELVEY (Somerset)
CHELWOOD (Somerset)
Cheney Longville (Shropshire)
Chepstow (Monmouthshire)
CHEQUERBENT (Greater Manchester)
Cherington (Gloucestershire)
Cherington (Warwickshire)
Cheriton (Kent)
Cheriton (Swansea)

See paras. 2 and 4 of the User Guide 2003 if you can't find your place name.

31

Cheriton or Stackpole Elidor (Pembrokeshire)
Cherrington (Shropshire)
Cherry Burton (East Riding of Yorkshire)
Cherry Orchard (Hereford & Worcester)
Cherry Willingham (Lincolnshire)
CHESHAM (Greater Manchester)
Chesley (Kent)
CHESLYN HAY (Staffordshire)
Chessetts Wood (Warwickshire)
Chester (Cheshire)
CHESTER MOOR (Durham)
CHESTER-LE-STREET (Durham)
Chesterblade (Somerset)
CHESTERFIELD (Derbyshire)
Chesterfield (Staffordshire)
Chesterton (Gloucestershire)
Chesterton (Shropshire)
CHESTERTON (Derbyshire)
Chesterton Green (Warwickshire)
CHESTERWOOD (Northumberland)
Chestfield (Kent)
Chestnut Street (Kent)
Cheswardine (Shropshire)
CHESWELL (Shropshire)
CHESWICK (Northumberland)
Cheswick Green (West Midlands)
CHETTON (Shropshire)
Chetwynd (Shropshire)
CHETWYND ASTON (Shropshire)
Chevening (Kent)
CHEVINGTON DRIFT (Northumberland)
CHEW MAGNA (Somerset)
CHEW MOOR (Greater Manchester)
Chew Stoke (Somerset)
Chewton Keynsham (Somerset)
Chewton Mendip (Somerset)
Chickward (Hereford & Worcester)
Chiddingstone (Kent)
Chiddingstone Causeway (Kent)
CHIDSWELL (West Yorkshire)
CHILCOMPTON (Somerset)
Chilcote (Leicestershire)
Child's Ercall (Shropshire)
Childer Thornton (Cheshire)
Childswickham (Hereford & Worcester)
Childwall (Merseyside)
Chilham (Kent)
CHILLENDEN (Kent)
CHILLINGHAM (Northumberland)
Chillington (Somerset)
Chilmington Green (Kent)
Chilthorne Domer (Somerset)
Chilton (Kent)
Chilton Cantelo (Somerset)
Chilton Polden (Somerset)
Chilton Trinity (Somerset)
CHILWELL (Nottinghamshire)
CHINLEY (Derbyshire)
CHIPCHASE CASTLE (Northumberland)
Chipnall (Shropshire)
Chipping (Lancashire)
Chipping Campden (Gloucestershire)
Chipping Sodbury (Gloucestershire)
Chipstable (Somerset)
Chipstead (Kent)
Chirbury (Shropshire)
CHIRK (Wrexham)
Chiselborough (Somerset)
CHISLET (Kent)
CHISLEY (West Yorkshire)
CHISWORTH (Derbyshire)
CHITTLEHAMPTON (Devon)
Chollerford (Northumberland)
Chollerton (Northumberland)
Cholstrey (Hereford & Worcester)
CHOP GATE (North Yorkshire)
CHOPPINGTON (Northumberland)
CHOPWELL (Tyne & Wear)
Chorley (Cheshire)
CHORLEY (Lancashire)
CHORLEY (Shropshire)
Chorley (Staffordshire)
Chorlton (Cheshire)
Chorlton Lane (Cheshire)
Chorlton-cum-Hardy (Greater Manchester)
Choulton (Shropshire)
Chowley (Cheshire)
CHRISTCHURCH (Gloucestershire)
Christleton (Cheshire)
Christon (Somerset)
CHRISTON BANK (Northumberland)

Chunal (Derbyshire)
CHURCH (Lancashire)
CHURCH ASHTON (Shropshire)
CHURCH BROUGH (Cumbria)
Church Broughton (Derbyshire)
Church Eaton (Staffordshire)
Church End (Lincolnshire)
CHURCH END (Warwickshire)
Church Fenton (North Yorkshire)
CHURCH GRESLEY (Derbyshire)
Church Hill (Cheshire)
CHURCH HILL (Staffordshire)
CHURCH HOUSES (North Yorkshire)
Church Laneham (Nottinghamshire)
Church Langton (Leicestershire)
Church Lawford (Warwickshire)
CHURCH LAWTON (Staffordshire)
Church Leigh (Staffordshire)
Church Lench (Hereford & Worcester)
Church Mayfield (Staffordshire)
Church Minshull (Cheshire)
Church Preen (Shropshire)
CHURCH PULVERBATCH (Shropshire)
Church Stoke (Powys)
Church Street (Kent)
Church Stretton (Shropshire)
Church Town (Lincolnshire)
CHURCH VILLAGE (Rhondda Cynon Taff)
CHURCH WARSOP (Nottinghamshire)
Church Wilne (Derbyshire)
Churcham (Gloucestershire)
CHURCHBRIDGE (Staffordshire)
Churchdown (Gloucestershire)
CHURCHFIELD (West Midlands)
Churchill (Hereford & Worcester)
Churchill (Somerset)
Churchover (Warwickshire)
Churchstanton (Somerset)
Churchthorpe (Lincolnshire)
Churchtown (Derbyshire)
Churchtown (Lancashire)
Churchtown (Merseyside)
Churnsike Lodge (Northumberland)
Churton (Cheshire)
CHURWELL (West Yorkshire)
Cil (Powys)
Cilcain (Flintshire)
Cilcewydd (Powys)
CILFREW (Neath Port Talbot)
CILFYNYDD (Rhondda Cynon Taff)
Cilgerran (Pembrokeshire)
Cilgwyn (Carmarthenshire)
CILMAENGWYN (Neath Port Talbot)
Cilmery (Powys)
Cilrhedyn (Pembrokeshire)
Cilsan (Carmarthenshire)
Cilycwm (Carmarthenshire)
CIMLA (Neath Port Talbot)
CINDER HILL (West Midlands)
CINDERFORD (Gloucestershire)
Cirencester (Gloucestershire)
Citadilla (North Yorkshire)
City (Vale of Glamorgan)
City Dulas (Anglesey)
Cladswell (Hereford & Worcester)
CLANDOWN (Somerset)
Clanville (Somerset)
Clap Hill (Kent)
Clapham (North Yorkshire)
CLAPWORTHY (Devon)
Clappersgate (Cumbria)
CLAPTON (Somerset)
CLAPTON-IN-GORDANO (Somerset)
Clapton-on-the-Hill (Gloucestershire)
CLARAVALE (Tyne & Wear)
Clarbeston (Pembrokeshire)
Clarbeston Road (Pembrokeshire)
CLARBOROUGH (Nottinghamshire)
CLAREWOOD (Northumberland)
Clatter (Powys)
Clatworthy (Somerset)
CLAUGHTON (Lancashire)
Claughton (Merseyside)
Clavelshay (Somerset)
Claverdon (Warwickshire)
Claverham (Somerset)
Claverley (Shropshire)
Claverton (Somerset)
Claverton Down (Somerset)
Clawdd-coch (Vale of Glamorgan)
Clawdd-newydd (Denbighshire)

Clawthorpe (Cumbria)
Claxby (Lincolnshire)
Claxton (North Yorkshire)
CLAY CROSS (Derbyshire)
CLAY HILL (Bristol)
Claybrooke Magna (Leicestershire)
Claygate (Kent)
Claygate Cross (Kent)
CLAYHANGER (West Midlands)
Claypits (Gloucestershire)
Claypole (Lincolnshire)
Claythorpe (Lincolnshire)
CLAYTON (South Yorkshire)
CLAYTON (West Yorkshire)
CLAYTON GREEN (Lancashire)
CLAYTON WEST (West Yorkshire)
CLAYTON-LE-MOORS (Lancashire)
CLAYTON-LE-WOODS (Lancashire)
CLAYWORTH (Nottinghamshire)
CLEADON (Tyne & Wear)
CLEARWELL (Gloucestershire)
CLEARWELL MEEND (Gloucestershire)
Cleasby (North Yorkshire)
Cleatlam (Durham)
CLEATOR (Cumbria)
CLEATOR MOOR (Cumbria)
CLECKHEATON (West Yorkshire)
CLEE ST. MARGARET (Shropshire)
Cleedownton (Shropshire)
CLEEHILL (Shropshire)
Cleestanton (Shropshire)
Cleethorpes (Lincolnshire)
CLEETON ST. MARY (Shropshire)
Cleeve (Somerset)
Cleeve Hill (Gloucestershire)
Cleeve Prior (Hereford & Worcester)
Clehonger (Hereford & Worcester)
Clement Street (Kent)
Clent (Hereford & Worcester)
CLEOBURY MORTIMER (Shropshire)
CLEOBURY NORTH (Shropshire)
Clevedon (Somerset)
Cleveleys (Lancashire)
Clevelode (Hereford & Worcester)
Clewer (Somerset)
Cliburn (Cumbria)
CLIFF (Warwickshire)
Cliffe (Durham)
Cliffe (Kent)
CLIFFE (Lancashire)
CLIFFE (North Yorkshire)
Cliffe Woods (Kent)
Clifford (Hereford & Worcester)
Clifford (West Yorkshire)
Clifford Chambers (Warwickshire)
Clifford's Mesne (Gloucestershire)
Cliffsend (Kent)
Clifton (Bristol)
CLIFTON (Cumbria)
Clifton (Derbyshire)
CLIFTON (Greater Manchester)
Clifton (Hereford & Worcester)
Clifton (Lancashire)
Clifton (North Yorkshire)
CLIFTON (Northumberland)
CLIFTON (Nottinghamshire)
CLIFTON (South Yorkshire)
CLIFTON (West Yorkshire)
CLIFTON CAMPVILLE (Staffordshire)
CLIFTON DYKES (Cumbria)
Clifton upon Dunsmore (Warwickshire)
Clifton upon Teme (Hereford & Worcester)
Clifton Wood (Bristol)
Cliftonville (Kent)
Clink (Somerset)
Clint (North Yorkshire)
CLIPSTON (Nottinghamshire)
CLIPSTONE (Nottinghamshire)
Clitheroe (Lancashire)
Clive (Shropshire)
Clixby (Lincolnshire)
Clocaenog (Denbighshire)
CLOCK FACE (Merseyside)
Cloddiau (Powys)
Clodock (Hereford & Worcester)
Cloford (Somerset)
Closworth (Somerset)
Clotton (Cheshire)
Cloudesley Bush (Warwickshire)
CLOUGH (Greater Manchester)
CLOUGH FOOT (West Yorkshire)

See paras. 2 and 4 of the User Guide 2003 if you can't find your place name.

CLOUGH HEAD (North Yorkshire)
Cloughton (North Yorkshire)
Cloughton Newlands (North Yorkshire)
CLOW BRIDGE (Lancashire)
CLOWNE (Derbyshire)
CLOWS TOP (Hereford & Worcester)
CLOY (Wrexham)
Clubmoor (Merseyside)
Clun (Shropshire)
Clunbury (Shropshire)
Clungunford (Shropshire)
Clunton (Shropshire)
Clutton (Cheshire)
CLUTTON (Somerset)
CLUTTON HILL (Somerset)
CLYDACH (Monmouthshire)
CLYDACH VALE (Rhondda Cynon Taff)
Clydey (Pembrokeshire)
Clynderwen (Carmarthenshire)
CLYNE (Neath Port Talbot)
Clyro (Powys)
COAL ASTON (Derbyshire)
COAL POOL (West Midlands)
COALBROOKDALE (Shropshire)
COALBROOKVALE (Blaenau Gwent)
COALBURNS (Tyne & Wear)
COALCLEUGH (Northumberland)
Coaley (Gloucestershire)
COALFELL (Cumbria)
COALMOOR (Shropshire)
COALPIT HEATH (Gloucestershire)
COALPIT HILL (Staffordshire)
COALPORT (Shropshire)
COALVILLE (Leicestershire)
COANWOOD (Northumberland)
Coat (Somerset)
Coates (Gloucestershire)
Coates (Lincolnshire)
Coatham (North Yorkshire)
Coatham Mundeville (Durham)
Coberley (Gloucestershire)
Cobhall Common (Hereford & Worcester)
Cobham (Kent)
Cobnash (Hereford & Worcester)
COBRIDGE (Staffordshire)
COCK ALLEY (Derbyshire)
COCK BANK (Wrexham)
Cock Bevington (Warwickshire)
Cock Street (Kent)
COCKAYNE (North Yorkshire)
Cocker Bar (Lancashire)
COCKER BROOK (Lancashire)
COCKERDALE (West Yorkshire)
Cockerham (Lancashire)
Cockermouth (Cumbria)
COCKETT (Swansea)
COCKFIELD (Durham)
Cocklake (Somerset)
COCKLE PARK (Northumberland)
Cockley Beck (Cumbria)
COCKSHUTFORD (Shropshire)
Cockshutt (Shropshire)
Cockwood (Somerset)
COCKYARD (Derbyshire)
Cockyard (Hereford & Worcester)
Coddington (Cheshire)
Coddington (Hereford & Worcester)
Coddington (Nottinghamshire)
CODNOR (Derbyshire)
Codrington (Gloucestershire)
Codsall (Staffordshire)
Codsall Wood (Staffordshire)
Coed Morgan (Monmouthshire)
COED TALON (Flintshire)
Coed-y-caerau (Newport)
Coed-y-paen (Monmouthshire)
Coed-yr-ynys (Powys)
Coedana (Anglesey)
COEDELY (Rhondda Cynon Taff)
Coedkernew (Newport)
COEDPOETH (Wrexham)
Coedway (Powys)
COELBREN (Powys)
Cofton Hackett (Hereford & Worcester)
Cogan (Vale of Glamorgan)
Coity (Bridgend)
Colburn (North Yorkshire)
Colby (Cumbria)
Cold Ashton (Gloucestershire)
Cold Aston (Gloucestershire)
Cold Blow (Pembrokeshire)

COLD COTES (North Yorkshire)
Cold Green (Hereford & Worcester)
Cold Hanworth (Lincolnshire)
Cold Hatton (Shropshire)
Cold Hatton Heath (Shropshire)
COLD HESLEDON (Durham)
COLD HIENDLEY (West Yorkshire)
COLD KIRBY (North Yorkshire)
Cold Newton (Leicestershire)
Cold Overton (Leicestershire)
Cold Weston (Shropshire)
COLDBECK (Cumbria)
Colden (West Yorkshire)
Coldharbour (Gloucestershire)
COLDMEECE (Staffordshire)
COLDRED (Kent)
Coldwell (Hereford & Worcester)
Cole (Somerset)
COLE END (Warwickshire)
Colebatch (Shropshire)
Coleby (Lincolnshire)
COLEFORD (Gloucestershire)
COLEFORD (Somerset)
Coleford Water (Somerset)
Colemere (Shropshire)
COLEMORE GREEN (Shropshire)
COLEORTON (Leicestershire)
Colesbourne (Gloucestershire)
COLESHILL (Warwickshire)
COLEY (Somerset)
College Green (Somerset)
Collier Street (Kent)
Collier's Green (Kent)
Colliers Green (Kent)
COLLIERY ROW (Tyne & Wear)
COLLINGHAM (Nottinghamshire)
Collingham (West Yorkshire)
Collington (Hereford & Worcester)
COLLINS GREEN (Cheshire)
Collins Green (Hereford & Worcester)
Coln Rogers (Gloucestershire)
Coln St. Aldwyns (Gloucestershire)
Coln St. Dennis (Gloucestershire)
COLNE (Lancashire)
COLNE BRIDGE (West Yorkshire)
COLNE EDGE (Lancashire)
COLSTERDALE (North Yorkshire)
Colsterworth (Lincolnshire)
COLSTON BASSETT (Nottinghamshire)
Colt's Hill (Kent)
Colton (Cumbria)
COLTON (North Yorkshire)
COLTON (Staffordshire)
COLTON (West Yorkshire)
Colva (Powys)
Colwall (Hereford & Worcester)
Colwell (Northumberland)
COLWICH (Staffordshire)
COLWICK (Nottinghamshire)
Colwinston (Vale of Glamorgan)
Combe (Hereford & Worcester)
Combe Florey (Somerset)
Combe Hay (Somerset)
Combe Moor (Hereford & Worcester)
Combe St. Nicholas (Somerset)
Comberbach (Cheshire)
COMBERFORD (Staffordshire)
Comberton (Hereford & Worcester)
Combridge (Staffordshire)
Combrook (Warwickshire)
COMBS (Derbyshire)
Combwich (Somerset)
Comhampton (Hereford & Worcester)
Commercial (Pembrokeshire)
Commins Coch (Powys)
Common Edge (Lancashire)
COMMON END (Cumbria)
COMMON SIDE (Derbyshire)
COMMONDALE (North Yorkshire)
Commonside (Cheshire)
Commonside (Derbyshire)
Commonwood (Shropshire)
COMMONWOOD (Wrexham)
Compass (Somerset)
COMPSTALL (Greater Manchester)
Compton (Staffordshire)
Compton Abdale (Gloucestershire)
Compton Bishop (Somerset)
COMPTON DANDO (Somerset)
Compton Dundon (Somerset)
Compton Durville (Somerset)

Compton Greenfield (Gloucestershire)
Compton Martin (Somerset)
Compton Pauncefoot (Somerset)
Compton Verney (Warwickshire)
Conder Green (Lancashire)
Conderton (Hereford & Worcester)
Condicote (Gloucestershire)
CONDOVER (Shropshire)
Coney Hill (Gloucestershire)
Coneysthorpe (North Yorkshire)
Congerstone (Leicestershire)
Conghurst (Kent)
Congleton (Cheshire)
Congresbury (Somerset)
Congreve (Staffordshire)
Coningsby (Lincolnshire)
CONISBROUGH (South Yorkshire)
Conisholme (Lincolnshire)
Coniston (Cumbria)
Coniston (East Riding of Yorkshire)
Coniston Cold (North Yorkshire)
CONISTONE (North Yorkshire)
CONNAH'S QUAY (Flintshire)
CONONLEY (North Yorkshire)
CONSALL (Staffordshire)
CONSETT (Durham)
Constable Burton (North Yorkshire)
CONSTABLE LEE (Lancashire)
Conyer (Kent)
Cookhill (Warwickshire)
Cookley (Hereford & Worcester)
Cooksey Green (Hereford & Worcester)
COOKSHILL (Staffordshire)
Cookson Green (Cheshire)
COOKSON'S GREEN (Durham)
Cooling (Kent)
Cooling Street (Kent)
Coombe (Gloucestershire)
Coombe Dingle (Bristol)
Coombe End (Somerset)
Coombe Hill (Gloucestershire)
Coombe Street (Somerset)
COOMBESWOOD (West Midlands)
Cooper Street (Kent)
COOPER TURNING (Greater Manchester)
Cooper's Corner (Kent)
Cop Street (Kent)
Copgrove (North Yorkshire)
COPLEY (Durham)
COPLEY (Greater Manchester)
Copley (West Yorkshire)
Coplow Dale (Derbyshire)
COPMANTHORPE (North Yorkshire)
Copmere End (Staffordshire)
Copp (Lancashire)
Coppenhall (Staffordshire)
Coppenhall Moss (Cheshire)
COPPICEGATE (Shropshire)
Coppins Corner (Kent)
COPPULL (Lancashire)
COPPULL MOOR (Lancashire)
Copster Green (Lancashire)
Copston Magna (Warwickshire)
Copt Heath (West Midlands)
Copt Hewick (North Yorkshire)
CORBRIDGE (Northumberland)
Corby Glen (Lincolnshire)
Corby Hill (Cumbria)
CORDWELL (Derbyshire)
CORELEY (Shropshire)
Corfe (Somerset)
Corfton (Shropshire)
Corks Pond (Kent)
CORLEY (Warwickshire)
CORLEY ASH (Warwickshire)
CORLEY MOOR (Warwickshire)
Corner Row (Lancashire)
Corney (Cumbria)
CORNFORTH (Durham)
Cornhill-on-Tweed (Northumberland)
CORNHOLME (West Yorkshire)
Cornriggs (Durham)
CORNSAY (Durham)
CORNSAY COLLIERY (Durham)
Corntown (Vale of Glamorgan)
Corringham (Lincolnshire)
Corse (Gloucestershire)
Corse Lawn (Gloucestershire)
Corston (Somerset)
Corton Denham (Somerset)
Corwen (Denbighshire)

See paras. 2 and 4 of the User Guide 2003 if you can't find your place name.

Cosby (Leicestershire)
COSELEY (West Midlands)
Cosford (Shropshire)
Cosheston (Pembrokeshire)
COSSALL (Nottinghamshire)
COSSALL MARSH (Nottinghamshire)
Cossington (Leicestershire)
Cossington (Somerset)
Costock (Nottinghamshire)
Coston (Leicestershire)
Cote (Somerset)
Cotebrook (Cheshire)
Cotehill (Cumbria)
Cotes (Cumbria)
Cotes (Leicestershire)
COTES (Staffordshire)
COTES HEATH (Staffordshire)
Cotesbach (Leicestershire)
COTGRAVE (Nottinghamshire)
Cotham (Bristol)
Cotham (Nottinghamshire)
Cothelstone (Somerset)
Cotherstone (Durham)
COTMANHAY (Derbyshire)
Coton (Shropshire)
COTON (Staffordshire)
Coton Clanford (Staffordshire)
Coton Hayes (Staffordshire)
COTON HILL (Shropshire)
Coton in the Clay (Staffordshire)
COTON IN THE ELMS (Derbyshire)
COTON PARK (Derbyshire)
Cottam (East Riding of Yorkshire)
Cottam (Lancashire)
Cottam (Nottinghamshire)
COTTERDALE (North Yorkshire)
Cotteridge (West Midlands)
Cottingham (East Riding of Yorkshire)
COTTINGLEY (West Yorkshire)
COTTON TREE (Lancashire)
Cottrell (Vale of Glamorgan)
Cotwall (Shropshire)
Cotwalton (Staffordshire)
COUGHTON (Hereford & Worcester)
Coughton (Warwickshire)
Coulderton (Cumbria)
Coultings (Somerset)
Coulton (North Yorkshire)
COUND (Shropshire)
COUNDLANE (Shropshire)
COUNDON (Durham)
COUNDON GRANGE (Durham)
COUNTERSETT (North Yorkshire)
Countesthorpe (Leicestershire)
Coup Green (Lancashire)
Coupland (Cumbria)
Coupland (Northumberland)
Court Henry (Carmarthenshire)
Court-at-Street (Kent)
Courtway (Somerset)
COVEN (Staffordshire)
COVEN LAWN (Staffordshire)
Covenham St. Bartholomew (Lincolnshire)
Covenham St. Mary (Lincolnshire)
COVENTRY (West Midlands)
COVERHAM (North Yorkshire)
Cow Honeybourne (Hereford & Worcester)
COWAN BRIDGE (Lancashire)
Cowbit (Lincolnshire)
Cowbridge (Vale of Glamorgan)
Cowdale (Derbyshire)
Cowden (Kent)
Cowden Pound (Kent)
Cowden Station (Kent)
Cowers Lane (Derbyshire)
Cowesby (North Yorkshire)
COWGILL (Cumbria)
Cowhill (Gloucestershire)
COWLEY (Derbyshire)
Cowley (Gloucestershire)
COWLING (Lancashire)
Cowling (North Yorkshire)
COWMES (West Yorkshire)
COWPE (Lancashire)
COWPEN (Northumberland)
Cowpen Bewley (Durham)
Cowshill (Durham)
Cowslip Green (Somerset)
Cowthorpe (North Yorkshire)
Coxall (Shropshire)
Coxbank (Cheshire)

COXBENCH (Derbyshire)
Coxbridge (Somerset)
Coxgreen (Staffordshire)
Coxheath (Kent)
COXHOE (Durham)
Coxley (Somerset)
COXLEY (West Yorkshire)
Coxley Wick (Somerset)
Coxwold (North Yorkshire)
Coychurch (Bridgend)
COYTRAHEN (Bridgend)
Crabbs Cross (Hereford & Worcester)
CRABTREE GREEN (Wrexham)
Crackenthorpe (Cumbria)
CRACKLEY (Staffordshire)
Crackley (Warwickshire)
CRACKLEYBANK (Shropshire)
Crackpot (North Yorkshire)
CRACOE (North Yorkshire)
Cradley (Hereford & Worcester)
CRADLEY (West Midlands)
Cradoc (Powys)
Crag Foot (Lancashire)
CRAGG HILL (West Yorkshire)
Cragg Vale (West Yorkshire)
CRAGHEAD (Durham)
Crai (Powys)
CRAIG LLANGIWG (Neath Port Talbot)
Craig Penllyn (Vale of Glamorgan)
CRAIG-Y-DUKE (Neath Port Talbot)
CRAIG-Y-NOS (Powys)
CRAIGCEFNPARC (Swansea)
Craignant (Shropshire)
Crakehall (North Yorkshire)
Crakehill (North Yorkshire)
Crakemarsh (Staffordshire)
Crambe (North Yorkshire)
CRAMLINGTON (Northumberland)
Cranage (Cheshire)
CRANBERRY (Staffordshire)
Cranbrook (Kent)
Cranbrook Common (Kent)
CRANE MOOR (South Yorkshire)
Cranham (Gloucestershire)
Cranhill (Warwickshire)
CRANK (Merseyside)
Cranmore (Somerset)
Cranoe (Leicestershire)
Cranswick (East Riding of Yorkshire)
Cranwell (Lincolnshire)
Craster (Northumberland)
Craswall (Hereford & Worcester)
Crateford (Staffordshire)
Crathorne (North Yorkshire)
Craven Arms (Shropshire)
CRAWCROOK (Tyne & Wear)
CRAWFORD (Lancashire)
CRAWLEY SIDE (Durham)
CRAWSHAWBOOTH (Lancashire)
Crawton (North Yorkshire)
CRAY (North Yorkshire)
Crayke (North Yorkshire)
Craythorne (Staffordshire)
Creamore Bank (Shropshire)
Credenhill (Hereford & Worcester)
Creech Heathfield (Somerset)
Creech St. Michael (Somerset)
Creeton (Lincolnshire)
Cregrina (Powys)
CREIGIAU (Cardiff)
Cressage (Shropshire)
Cressbrook (Derbyshire)
CRESSELLY (Pembrokeshire)
CRESSWELL (Northumberland)
CRESSWELL (Pembrokeshire)
Cresswell (Staffordshire)
CRESWELL (Derbyshire)
CRESWELL GREEN (Staffordshire)
Crew Green (Powys)
Crewe (Cheshire)
Crewe Green (Cheshire)
Crewkerne (Somerset)
Crews Hill (Hereford & Worcester)
Crewton (Derbyshire)
Cribbs Causeway (Gloucestershire)
CRICH (Derbyshire)
Crich Carr (Derbyshire)
CRICH COMMON (Derbyshire)
Crick (Monmouthshire)
Crickadarn (Powys)
Cricket St. Thomas (Somerset)

Crickheath (Shropshire)
Crickhowell (Powys)
CRIDLING STUBBS (North Yorkshire)
Criggion (Powys)
CRIGGLESTONE (West Yorkshire)
CRIMBLE (Greater Manchester)
Crimscote (Warwickshire)
CRINGLES (North Yorkshire)
Crinow (Pembrokeshire)
Crizeley (Hereford & Worcester)
Crock Street (Somerset)
Crockenhill (Kent)
Crocker's Ash (Hereford & Worcester)
CROCKEY HILL (North Yorkshire)
Crockham Hill (Kent)
Crockhurst Street (Kent)
Croes-goch (Pembrokeshire)
Croes-y-mwyalch (Torfaen)
Croes-y-pant (Monmouthshire)
CROESERW (Neath Port Talbot)
Croesyceiliog (Carmarthenshire)
Croesyceiliog (Torfaen)
CROFT (Cheshire)
Croft (Leicestershire)
Croft (Lincolnshire)
Croft-on-Tees (Durham)
Crofton (Cumbria)
CROFTON (West Yorkshire)
Crofts Bank (Greater Manchester)
CROFTY (Swansea)
CROGLIN (Cumbria)
Cromford (Derbyshire)
CROMHALL (Gloucestershire)
CROMHALL COMMON (Gloucestershire)
CROMPTON FOLD (Greater Manchester)
CROMWELL (Nottinghamshire)
CRONTON (Merseyside)
Crook (Cumbria)
CROOK (Durham)
CROOKDAKE (Cumbria)
CROOKE (Greater Manchester)
CROOKED END (Gloucestershire)
Crooked Holme (Cumbria)
CROOKES (South Yorkshire)
CROOKHALL (Durham)
Crookham (Northumberland)
Crooklands (Cumbria)
Cropper (Derbyshire)
Cropston (Leicestershire)
Cropthorne (Hereford & Worcester)
Cropton (North Yorkshire)
CROPWELL BISHOP (Nottinghamshire)
CROPWELL BUTLER (Nottinghamshire)
CROSBY (Cumbria)
Crosby (Lincolnshire)
Crosby (Merseyside)
CROSBY GARRET (Cumbria)
CROSBY RAVENSWORTH (Cumbria)
CROSBY VILLA (Cumbria)
Croscombe (Somerset)
Crosemere (Shropshire)
Cross (Somerset)
Cross Ash (Monmouthshire)
CROSS FLATTS (West Yorkshire)
CROSS GATES (West Yorkshire)
CROSS GREEN (Staffordshire)
CROSS HANDS (Carmarthenshire)
CROSS HANDS (Pembrokeshire)
CROSS HILL (Derbyshire)
CROSS HILLS (North Yorkshire)
CROSS HOUSES (Shropshire)
Cross Inn (Pembrokeshire)
CROSS INN (Rhondda Cynon Taff)
CROSS LANE HEAD (Shropshire)
Cross Lanes (North Yorkshire)
CROSS LANES (Wrexham)
Cross o' th' hands (Derbyshire)
Cross Oak (Powys)
Cross Roads (Powys)
Cross Town (Cheshire)
Cross-at-Hand (Kent)
CROSSCANONBY (Cumbria)
Crossens (Merseyside)
Crossgate (Lincolnshire)
CROSSGATE (Staffordshire)
Crossgates (North Yorkshire)
Crossgates (Powys)
CROSSGILL (Lancashire)
Crosshands (Carmarthenshire)
Crosskeys (Caerphilly)
CROSSLAND EDGE (West Yorkshire)

See paras. 2 and 4 of the User Guide 2003 if you can't find your place name.

CROSSLAND HILL (West Yorkshire)
Crosslands (Cumbria)
Crosslanes (Shropshire)
CROSSLEY (West Yorkshire)
Crossway (Monmouthshire)
Crossway (Pembrokeshire)
Crossway (Powys)
Crossway Green (Hereford & Worcester)
Crossway Green (Monmouthshire)
Crosswell (Pembrokeshire)
Crosthwaite (Cumbria)
Croston (Lancashire)
Crouch (Kent)
Crough House Green (Kent)
CROW EDGE (South Yorkshire)
Crow Hill (Hereford & Worcester)
Crowcombe (Somerset)
Crowdecote (Derbyshire)
Crowden (Derbyshire)
Crowdleham (Kent)
CROWHOLE (Derbyshire)
Crowland (Lincolnshire)
Crowle (Hereford & Worcester)
CROWLE (Lincolnshire)
Crowle Green (Hereford & Worcester)
Crowsnest (Shropshire)
Crowton (Cheshire)
Croxall (Staffordshire)
Croxby (Lincolnshire)
CROXDALE (Durham)
Croxden (Staffordshire)
Croxteth (Merseyside)
Croxteth Park (Merseyside)
Croxton (Lincolnshire)
Croxton (Staffordshire)
Croxton Green (Cheshire)
CROXTON KERRIAL (Leicestershire)
Croxtonbank (Staffordshire)
CRUCKMEOLE (Shropshire)
CRUCKTON (Shropshire)
Crudgington (Shropshire)
Crug (Powys)
Crug-y-byddar (Powys)
Crugybar (Carmarthenshire)
CRUMLIN (Caerphilly)
CRUMPSALL (Greater Manchester)
Crundale (Kent)
Crundale (Pembrokeshire)
Crunwear (Pembrokeshire)
CRWBIN (Carmarthenshire)
Crymmych (Pembrokeshire)
CRYNANT (Neath Port Talbot)
Cubbington (Warwickshire)
CUBLEY (South Yorkshire)
Cublington (Hereford & Worcester)
Cucklington (Somerset)
CUCKNEY (Nottinghamshire)
Cuckold's Green (Kent)
Cuckoo Bridge (Lincolnshire)
Cuckoo's Nest (Cheshire)
Cuddington (Cheshire)
Cuddington Heath (Cheshire)
Cuddy Hill (Lancashire)
Cudworth (Somerset)
CUDWORTH (South Yorkshire)
CUDWORTH COMMON (South Yorkshire)
Cuerden Green (Lancashire)
CUERDLEY CROSS (Cheshire)
Culbone (Somerset)
CULCHETH (Cheshire)
Culgaith (Cumbria)
Culkerton (Gloucestershire)
CULLERCOATS (Tyne & Wear)
CULLINGWORTH (West Yorkshire)
Culmington (Shropshire)
Culverstone Green (Kent)
Culverthorpe (Lincolnshire)
Cumberworth (Lincolnshire)
Cumdivock (Cumbria)
Cummersdale (Cumbria)
CUMREW (Cumbria)
Cumwhinton (Cumbria)
Cumwhitton (Cumbria)
Cundall (North Yorkshire)
Curbar (Derbyshire)
Curdworth (Warwickshire)
Curland (Somerset)
Curry Mallet (Somerset)
Curry Rivel (Somerset)
Curteis Corner (Kent)
Curtisden Green (Kent)

Cushuish (Somerset)
Cusop (Hereford & Worcester)
Cutcombe (Somerset)
CUTGATE (Greater Manchester)
Cutnall Green (Hereford & Worcester)
Cutsdean (Gloucestershire)
CUTSYKE (West Yorkshire)
CUTTHORPE (Derbyshire)
Cuxton (Kent)
Cuxwold (Lincolnshire)
CWM (Blaenau Gwent)
Cwm (Denbighshire)
CWM CAPEL (Carmarthenshire)
CWM DULAIS (Swansea)
Cwm Irfon (Powys)
Cwm Morgan (Carmarthenshire)
CWM-BACH (Carmarthenshire)
CWM-CELYN (Blaenau Gwent)
Cwm-cou (Carmarthenshire)
Cwm-Crownon (Powys)
Cwm-Llinau (Powys)
CWM-Y-GLO (Carmarthenshire)
CWMAFAN (Neath Port Talbot)
CWMAMAN (Rhondda Cynon Taff)
Cwmann (Carmarthenshire)
CWMAVON (Torfaen)
Cwmbach (Carmarthenshire)
Cwmbach (Powys)
CWMBACH (Rhondda Cynon Taff)
Cwmbach Llechrhyd (Powys)
Cwmbelan (Powys)
CWMBRAN (Torfaen)
CWMCARN (Caerphilly)
Cwmcarvan (Monmouthshire)
CWMDARE (Rhondda Cynon Taff)
Cwmdu (Carmarthenshire)
Cwmdu (Powys)
CWMDU (Swansea)
Cwmduad (Carmarthenshire)
Cwmdwr (Carmarthenshire)
CWMFELIN (Merthyr Tydfil)
Cwmfelin Boeth (Carmarthenshire)
Cwmfelin Mynach (Carmarthenshire)
CWMFELINFACH (Caerphilly)
Cwmffrwd (Carmarthenshire)
CWMGIEDD (Powys)
CWMGORSE (Carmarthenshire)
CWMGWILI (Carmarthenshire)
CWMGWRACH (Neath Port Talbot)
Cwmhiraeth (Carmarthenshire)
Cwmisfael (Carmarthenshire)
CWMLLYNFELL (Neath Port Talbot)
CWMPARC (Rhondda Cynon Taff)
Cwmpengraig (Carmarthenshire)
CWMPENNAR (Rhondda Cynon Taff)
Cwmrhos (Powys)
CWMRHYDYCEIRW (Swansea)
CWMTILLERY (Blaenau Gwent)
Cwmyoy (Monmouthshire)
Cwrt-y-gollen (Powys)
Cyfronydd (Powys)
CYLIBEBYLL (Neath Port Talbot)
CYMER (Neath Port Talbot)
CYMMER (Rhondda Cynon Taff)
Cynghordy (Carmarthenshire)
CYNHEIDRE (Carmarthenshire)
CYNONVILLE (Neath Port Talbot)
Cynwyd (Denbighshire)
Cynwyl Elfed (Carmarthenshire)

D

Dacre (Cumbria)
DACRE (North Yorkshire)
DACRE BANKS (North Yorkshire)
Daddry Shield (Durham)
Dadlington (Leicestershire)
DAFEN (Carmarthenshire)
Daglingworth (Gloucestershire)
DAISY HILL (Greater Manchester)
DAISY HILL (West Yorkshire)
Dalbury (Derbyshire)
Dalby (Lincolnshire)
DALBY (North Yorkshire)
Dalderby (Lincolnshire)
Dale (Cumbria)
DALE (Derbyshire)
Dale (Pembrokeshire)
Dale Bottom (Cumbria)
Dale End (Derbyshire)
Dale End (North Yorkshire)

DALEHOUSE (North Yorkshire)
DALLOW (North Yorkshire)
Dalston (Cumbria)
Dalton (Cumbria)
DALTON (Lancashire)
Dalton (North Yorkshire)
Dalton (Northumberland)
DALTON (South Yorkshire)
DALTON MAGNA (South Yorkshire)
DALTON PARVA (South Yorkshire)
Dalton Piercy (Durham)
Dalton-in-Furness (Cumbria)
DALTON-LE-DALE (Durham)
Dalton-on-Tees (North Yorkshire)
DAN'S CASTLE (Durham)
Dan-y-Parc (Powys)
Danaway (Kent)
DANBY (North Yorkshire)
DANBY BOTTOM (North Yorkshire)
Danby Wiske (North Yorkshire)
Dane Hills (Leicestershire)
Dane Street (Kent)
DANEBRIDGE (Cheshire)
Danesford (Shropshire)
DANESMOOR (Derbyshire)
Daniel's Water (Kent)
Danthorpe (East Riding of Yorkshire)
Danzey Green (Warwickshire)
Dapple Heath (Staffordshire)
DARCY LEVER (Greater Manchester)
DAREN-FELEN (Monmouthshire)
Darenth (Kent)
Daresbury (Cheshire)
DARFIELD (South Yorkshire)
Dargate (Kent)
Darland (Kent)
DARLAND (Wrexham)
DARLASTON (Staffordshire)
DARLASTON (West Midlands)
DARLASTON GREEN (West Midlands)
DARLEY (North Yorkshire)
Darley Abbey (Derbyshire)
Darley Bridge (Derbyshire)
Darley Dale (Derbyshire)
Darley Green (Warwickshire)
DARLEY HEAD (North Yorkshire)
Darlingscott (Warwickshire)
Darlington (Durham)
Darliston (Shropshire)
DARLTON (Nottinghamshire)
Darnford (Staffordshire)
Darowen (Powys)
DARRAS HALL (Northumberland)
DARRINGTON (West Yorkshire)
Darshill (Somerset)
Dartford (Kent)
DARTON (South Yorkshire)
DARWEN (Lancashire)
DAUBHILL (Greater Manchester)
Davenham (Cheshire)
Davenport (Greater Manchester)
Davenport Green (Cheshire)
Davenport Green (Greater Manchester)
David Street (Kent)
Davington Hill (Kent)
Davyhulme (Greater Manchester)
DAW END (West Midlands)
DAWLEY (Shropshire)
Daws Green (Somerset)
Dawsmere (Lincolnshire)
Day Green (Cheshire)
DAYBROOK (Nottinghamshire)
Dayhills (Staffordshire)
Dayhouse Bank (Hereford & Worcester)
Daylesford (Gloucestershire)
Ddol (Flintshire)
Ddol-Cownwy (Powys)
DEAL (Kent)
DEAN (Cumbria)
DEAN (Lancashire)
Dean (Somerset)
Dean Bottom (Kent)
DEAN HEAD (South Yorkshire)
Dean Row (Cheshire)
Dean Street (Kent)
DEANE (Greater Manchester)
Deanhead (West Yorkshire)
DEANRAW (Northumberland)
DEANSCALES (Cumbria)
DEARHAM (Cumbria)
DEARNLEY (Greater Manchester)

See paras. 2 and 4 of the User Guide 2003 if you can't find your place name.

Deblin's Green (Hereford & Worcester)
DEEPCAR (South Yorkshire)
DEEPDALE (Cumbria)
Deepdale (North Yorkshire)
Deeping Gate (Lincolnshire)
Deeping St. James (Lincolnshire)
Deeping St. Nicholas (Lincolnshire)
Deerhurst (Gloucestershire)
Deerhurst Walton (Gloucestershire)
DEERLAND (Pembrokeshire)
Deerton Street (Kent)
Defford (Hereford & Worcester)
Defynnog (Powys)
DEIGHTON (North Yorkshire)
DEIGHTON (West Yorkshire)
Delamere (Cheshire)
Delmonden Green (Kent)
DELPH (Greater Manchester)
DELVES (Durham)
Dembleby (Lincolnshire)
DENABY (South Yorkshire)
DENABY MAIN (South Yorkshire)
Denbigh (Denbighshire)
DENBY (Derbyshire)
DENBY BOTTLES (Derbyshire)
DENBY DALE (West Yorkshire)
Dendron (Cumbria)
DENHOLME (West Yorkshire)
DENHOLME CLOUGH (West Yorkshire)
DENSHAW (Greater Manchester)
Densole (Kent)
Denstone (Staffordshire)
Denstroude (Kent)
Dent (Cumbria)
Dent-de-Lion (Kent)
Denton (Durham)
DENTON (Greater Manchester)
DENTON (Kent)
Denton (Lincolnshire)
Denton (North Yorkshire)
DENWICK (Northumberland)
Derby (Derbyshire)
DERI (Caerphilly)
Derringstone (Kent)
Derrington (Staffordshire)
Derrythorpe (Lincolnshire)
Derwen (Denbighshire)
Derwen Fawr (Carmarthenshire)
Derwenlas (Powys)
Derwydd (Carmarthenshire)
DESFORD (Leicestershire)
DETCHANT (Northumberland)
Detling (Kent)
DEUXHILL (Shropshire)
Devauden (Monmouthshire)
DEVITTS GREEN (Warwickshire)
DEWSBURY (West Yorkshire)
DEWSBURY MOOR (West Yorkshire)
Deytheur (Powys)
Dial (Somerset)
Dickens Heath (West Midlands)
Didbrook (Gloucestershire)
Diddlebury (Shropshire)
Didley (Hereford & Worcester)
Didmarton (Gloucestershire)
Didsbury (Greater Manchester)
Digby (Lincolnshire)
Diggle (Greater Manchester)
DIGMOOR (Lancashire)
DIGMORE (Lancashire)
DILHORNE (Staffordshire)
Dilston (Northumberland)
Dilwyn (Hereford & Worcester)
DIMPLE (Derbyshire)
DIMPLE (Greater Manchester)
Dinas (Carmarthenshire)
Dinas (Pembrokeshire)
DINAS (Rhondda Cynon Taff)
Dinas Powys (Vale of Glamorgan)
Dinder (Somerset)
Dinedor (Hereford & Worcester)
Dingestow (Monmouthshire)
Dingle (Merseyside)
Dingleden (Kent)
Dinham (Monmouthshire)
Dinnington (Somerset)
DINNINGTON (South Yorkshire)
DINNINGTON (Tyne & Wear)
Dipford (Somerset)
DIPTON (Durham)
Diptonmill (Northumberland)

Dirt Pot (Northumberland)
Discoed (Powys)
Diseworth (Leicestershire)
Dishforth (North Yorkshire)
DISLEY (Cheshire)
Disserth (Powys)
DISTINGTON (Cumbria)
DITCHBURN (Northumberland)
Ditcheat (Somerset)
DITHERINGTON (Shropshire)
DITTON (Cheshire)
Ditton (Kent)
Ditton Priors (Shropshire)
Dixton (Gloucestershire)
Dixton (Monmouthshire)
DOBCROSS (Greater Manchester)
Dobroyd Castle (West Yorkshire)
Docker (Lancashire)
Docklow (Hereford & Worcester)
Dockray (Cumbria)
Dod's Leigh (Staffordshire)
Dodd's Green (Cheshire)
Doddington (Kent)
Doddington (Lincolnshire)
Doddington (Northumberland)
DODDINGTON (Shropshire)
Dodford (Hereford & Worcester)
Dodington (Gloucestershire)
Dodington (Somerset)
Dodleston (Cheshire)
DODWORTH (South Yorkshire)
DODWORTH BOTTOM (South Yorkshire)
DODWORTH GREEN (South Yorkshire)
Doe Bank (West Midlands)
DOE LEA (Derbyshire)
Dogdyke (Lincolnshire)
DOGLEY LANE (West Yorkshire)
Dol-for (Powys)
Dol-gran (Carmarthenshire)
Dolancothi (Carmarthenshire)
Dolanog (Powys)
Dolau (Powys)
Doley (Shropshire)
Dolfor (Powys)
Dolley Green (Powys)
Dolphin (Flintshire)
Dolphinholme (Lancashire)
Dolyhir (Powys)
Domgay (Powys)
Donaldson's Lodge (Northumberland)
DONCASTER (South Yorkshire)
DONCASTER CARR (South Yorkshire)
Doniford (Somerset)
Donington (Lincolnshire)
Donington on Bain (Lincolnshire)
Donington Southing (Lincolnshire)
DONISTHORPE (Leicestershire)
Donkey Street (Kent)
Donnington (Gloucestershire)
Donnington (Hereford & Worcester)
DONNINGTON (Shropshire)
DONNINGTON WOOD (Shropshire)
Donyatt (Somerset)
DORDON (Warwickshire)
DORE (South Yorkshire)
Dormington (Hereford & Worcester)
Dormston (Hereford & Worcester)
Dorn (Gloucestershire)
Dorridge (West Midlands)
Dorrington (Lincolnshire)
DORRINGTON (Shropshire)
Dorsington (Warwickshire)
Dorstone (Hereford & Worcester)
DOSTHILL (Staffordshire)
Doughton (Gloucestershire)
Doulting (Somerset)
Dovaston (Shropshire)
DOVE GREEN (Nottinghamshire)
Dove Holes (Derbyshire)
DOVENBY (Cumbria)
DOVER (Greater Manchester)
Dover (Kent)
Doverdale (Hereford & Worcester)
Doveridge (Derbyshire)
Dowbridge (Lancashire)
Dowdeswell (Gloucestershire)
DOWLAIS (Merthyr Tydfil)
Dowlish Ford (Somerset)
Dowlish Wake (Somerset)
Down Ampney (Gloucestershire)
Down Hatherley (Gloucestershire)

DOWNEND (Gloucestershire)
Downham (Lancashire)
Downham (Northumberland)
Downhead (Somerset)
Downholland Cross (Lancashire)
Downholme (North Yorkshire)
Downing (Flintshire)
DOWNSIDE (Somerset)
Downton on the Rock (Hereford & Worcester)
Dowsby (Lincolnshire)
Dowsdale (Lincolnshire)
Doxey (Staffordshire)
DOXFORD (Northumberland)
Doynton (Gloucestershire)
DRAETHEN (Caerphilly)
Dragonby (Lincolnshire)
DRAKEHOLES (Nottinghamshire)
Drakelow (Hereford & Worcester)
Drakes Broughton (Hereford & Worcester)
Drakes Cross (Hereford & Worcester)
DRAUGHTON (North Yorkshire)
DRAX (North Yorkshire)
DRAX HALES (North Yorkshire)
Draycote (Warwickshire)
Draycott (Derbyshire)
Draycott (Gloucestershire)
Draycott (Hereford & Worcester)
Draycott (Shropshire)
Draycott (Somerset)
Draycott in the Clay (Staffordshire)
DRAYCOTT IN THE MOORS (Staffordshire)
Drayton (Hereford & Worcester)
Drayton (Leicestershire)
Drayton (Lincolnshire)
Drayton (Somerset)
DRAYTON BASSETT (Staffordshire)
DREBLEY (North Yorkshire)
Dreen Hill (Pembrokeshire)
DREENHILL (Pembrokeshire)
DREFACH (Carmarthenshire)
Drefelin (Carmarthenshire)
Drellingore (Kent)
DRESDEN (Staffordshire)
Driby (Lincolnshire)
Driffield (East Riding of Yorkshire)
Driffield (Gloucestershire)
Driffield Cross Roads (Gloucestershire)
Drigg (Cumbria)
DRIGHLINGTON (West Yorkshire)
Dringhoe (East Riding of Yorkshire)
Dringhouses (North Yorkshire)
Drinsey Nook (Nottinghamshire)
Drointon (Staffordshire)
Droitwich (Hereford & Worcester)
DRONFIELD (Derbyshire)
DRONFIELD WOODHOUSE (Derbyshire)
DROPPING WELL (South Yorkshire)
DROYLSDEN (Greater Manchester)
Druid (Denbighshire)
DRUIDS HEATH (West Midlands)
DRUIDSTON (Pembrokeshire)
Drumburgh (Cumbria)
Drumleaning (Cumbria)
DRURIDGE (Northumberland)
DRURY (Flintshire)
Dry Doddington (Lincolnshire)
DRYBECK (Cumbria)
DRYBROOK (Gloucestershire)
Dryslwyn (Carmarthenshire)
Dryton (Shropshire)
Duckington (Cheshire)
Duddlestone (Somerset)
Duddlewick (Shropshire)
DUDDO (Northumberland)
Duddon (Cheshire)
Duddon Bridge (Cumbria)
DUDLESTON (Shropshire)
Dudleston Heath (Shropshire)
DUDLEY (Tyne & Wear)
DUDLEY (West Midlands)
DUDLEY HILL (West Yorkshire)
DUDLEY PORT (West Midlands)
DUDNILL (Shropshire)
Duffield (Derbyshire)
DUFFRYN (Neath Port Talbot)
Dufton (Cumbria)
Duggleby (North Yorkshire)
DUKESTOWN (Blaenau Gwent)
DUKINFIELD (Greater Manchester)
Dulas (Anglesey)
Dulcote (Somerset)

Dulverton (Somerset)
Dumbleton (Gloucestershire)
Dumpton (Kent)
Dunball (Somerset)
Dunchurch (Warwickshire)
Dundon (Somerset)
Dundraw (Cumbria)
Dundry (Bristol)
Dundry (Somerset)
Dunfield (Gloucestershire)
DUNFORD BRIDGE (South Yorkshire)
Dungate (Kent)
DUNGWORTH (South Yorkshire)
DUNHAM (Nottinghamshire)
Dunham Town (Greater Manchester)
Dunham Woodhouses (Greater Manchester)
Dunham-on-the-Hill (Cheshire)
Dunhampstead (Hereford & Worcester)
Dunhampton (Hereford & Worcester)
Dunholme (Lincolnshire)
Dunk's Green (Kent)
DUNKERTON (Somerset)
Dunkeswick (West Yorkshire)
Dunkirk (Cheshire)
Dunkirk (Gloucestershire)
Dunkirk (Kent)
DUNKIRK (Staffordshire)
DUNLEY (Hereford & Worcester)
Dunn Street (Kent)
Dunnington (East Riding of Yorkshire)
Dunnington (North Yorkshire)
Dunnington (Warwickshire)
DUNNOCKSHAW (Lancashire)
Dunsa (Derbyshire)
Dunsby (Lincolnshire)
DUNSCAR (Greater Manchester)
DUNSCROFT (South Yorkshire)
Dunsdale (North Yorkshire)
DUNSILL (Nottinghamshire)
DUNSLEY (North Yorkshire)
Dunsley (Staffordshire)
Dunsop Bridge (Lancashire)
Dunstall (Staffordshire)
Dunstall Common (Hereford & Worcester)
DUNSTAN (Northumberland)
Dunstan Steads (Northumberland)
Dunster (Somerset)
Dunston (Lincolnshire)
Dunston (Staffordshire)
DUNSTON (Tyne & Wear)
Dunston Heath (Staffordshire)
DUNSVILLE (South Yorkshire)
Dunswell (East Riding of Yorkshire)
Duntisbourne Abbots (Gloucestershire)
Duntisbourne Rouse (Gloucestershire)
Dunton Bassett (Leicestershire)
Dunton Green (Kent)
DUNVANT (Swansea)
Dunwood (Staffordshire)
Durdar (Cumbria)
DURHAM (Durham)
DURKAR (West Yorkshire)
Durleigh (Somerset)
Durlock (Kent)
Durlow Common (Gloucestershire)
DURN (Greater Manchester)
Dursley (Gloucestershire)
Dursley Cross (Gloucestershire)
Durston (Somerset)
Dutlas (Powys)
Dutton (Cheshire)
Dwyran (Anglesey)
Dye House (Northumberland)
DYFATTY (Carmarthenshire)
Dyffryn (Anglesey)
DYFFRYN (Bridgend)
DYFFRYN (Merthyr Tydfil)
Dyffryn (Vale of Glamorgan)
DYFFRYN CELLWEN (Neath Port Talbot)
Dyke (Lincolnshire)
Dylife (Powys)
Dymchurch (Kent)
DYMOCK (Gloucestershire)
Dyrham (Gloucestershire)
Dyserth (Denbighshire)

E

Eachway (Hereford & Worcester)
Eachwick (Northumberland)
Eagland Hill (Lancashire)

EAGLE (Lincolnshire)
EAGLE BARNSDALE (Lincolnshire)
Eagle Manor (Lincolnshire)
Eaglescliffe (Durham)
EAGLESFIELD (Cumbria)
EAGLEY (Greater Manchester)
EAKRING (Nottinghamshire)
EALAND (Lincolnshire)
Eals (Northumberland)
EAMONT BRIDGE (Cumbria)
Earby (Lancashire)
EARCROFT (Lancashire)
EARDINGTON (Shropshire)
Eardisland (Hereford & Worcester)
Eardisley (Hereford & Worcester)
EARDISTON (Hereford & Worcester)
Eardiston (Shropshire)
Earl Shilton (Leicestershire)
Earl Sterndale (Derbyshire)
Earl's Croome (Hereford & Worcester)
Earle (Northumberland)
EARLESTOWN (Merseyside)
Earls Common (Hereford & Worcester)
EARLSDITTON (Shropshire)
EARLSDON (West Midlands)
EARLSHEATON (West Yorkshire)
Earlswood (Warwickshire)
Earlswood Common (Monmouthshire)
Earnshaw Bridge (Lancashire)
EARSDON (Northumberland)
EARSDON (Tyne & Wear)
Earswick (North Yorkshire)
Easby (North Yorkshire)
Easenhall (Warwickshire)
EASINGTON (Durham)
Easington (East Riding of Yorkshire)
Easington (North Yorkshire)
Easington (Northumberland)
EASINGTON COLLIERY (Durham)
EASINGTON LANE (Tyne & Wear)
Easingwold (North Yorkshire)
EASOLE STREET (Kent)
East Aberthaw (Vale of Glamorgan)
East Appleton (North Yorkshire)
East Ayton (North Yorkshire)
EAST BANK (Blaenau Gwent)
East Barkwith (Lincolnshire)
East Barming (Kent)
EAST BARNBY (North Yorkshire)
EAST BIERLEY (West Yorkshire)
EAST BOLDON (Tyne & Wear)
East Bolton (Northumberland)
East Bower (Somerset)
East Brent (Somerset)
EAST BRIDGFORD (Nottinghamshire)
EAST BRISCOE (Durham)
EAST BUTSFIELD (Durham)
East Butterwick (Lincolnshire)
East Carlton (West Yorkshire)
EAST CHEVINGTON (Northumberland)
East Chinnock (Somerset)
East Clevedon (Somerset)
East Coker (Somerset)
East Combe (Somerset)
East Compton (Somerset)
East Cote (Cumbria)
EAST COTTINGWITH (East Riding of Yorkshire)
EAST COWICK (East Riding of Yorkshire)
East Cowton (North Yorkshire)
EAST CRAMLINGTON (Northumberland)
East Cranmore (Somerset)
East Curthwaite (Cumbria)
East Dean (Gloucestershire)
East Drayton (Nottinghamshire)
East Dundry (Somerset)
East Ella (East Riding of Yorkshire)
East End (East Riding of Yorkshire)
East End (Kent)
EAST END (Somerset)
East Farleigh (Kent)
East Ferry (Lincolnshire)
East Firsby (Lincolnshire)
EAST GARFORTH (West Yorkshire)
East Goscote (Leicestershire)
East Halton (Lincolnshire)
EAST HARDWICK (West Yorkshire)
East Harlsey (North Yorkshire)
East Harptree (Somerset)
East Hartburn (Durham)
EAST HARTFORD (Northumberland)
East Hauxwell (North Yorkshire)

East Heckington (Lincolnshire)
EAST HEDLEYHOPE (Durham)
East Heslerton (North Yorkshire)
East Hewish (Somerset)
EAST HOLYWELL (Tyne & Wear)
East Horrington (Somerset)
EAST HORTON (Northumberland)
East Huntington (North Yorkshire)
East Huntspill (Somerset)
East Keal (Lincolnshire)
East Keswick (West Yorkshire)
East Kirkby (Lincolnshire)
EAST KYLOE (Northumberland)
East Lambrook (Somerset)
EAST LANGDON (Kent)
East Langton (Leicestershire)
EAST LAYTON (North Yorkshire)
East Leake (Nottinghamshire)
East Learmouth (Northumberland)
East Lound (Lincolnshire)
East Lutton (North Yorkshire)
East Lydeard (Somerset)
East Lydford (Somerset)
East Malling (Kent)
East Malling Heath (Kent)
EAST MARKHAM (Nottinghamshire)
East Marton (North Yorkshire)
EAST MORTON (West Yorkshire)
East Ness (North Yorkshire)
East Newton (East Riding of Yorkshire)
East Norton (Leicestershire)
EAST ORD (Northumberland)
East Peckham (Kent)
East Pennar (Pembrokeshire)
East Pennard (Somerset)
East Quantoxhead (Somerset)
East Rainham (Kent)
EAST RAINTON (Tyne & Wear)
East Ravendale (Lincolnshire)
East Rigton (West Yorkshire)
East Rolstone (Somerset)
East Rounton (North Yorkshire)
EAST SCRAFTON (North Yorkshire)
EAST SLEEKBURN (Northumberland)
East Stockwith (Lincolnshire)
East Stoke (Nottinghamshire)
EAST STOURMOUTH (Kent)
EAST STUDDAL (Kent)
East Sutton (Kent)
EAST-THE-WATER (Devon)
EAST THIRSTON (Northumberland)
East Torrington (Lincolnshire)
East Wall (Shropshire)
East Water (Somerset)
EAST WILLIAMSTON (Pembrokeshire)
EAST WITTON (North Yorkshire)
EAST WOODBURN (Northumberland)
East Woodlands (Somerset)
Eastbourne (Durham)
Eastbrook (Vale of Glamorgan)
Eastburn (West Yorkshire)
EASTBY (North Yorkshire)
Eastchurch (Kent)
Eastcombe (Gloucestershire)
Eastcote (West Midlands)
Easter Compton (Gloucestershire)
EASTERN GREEN (West Midlands)
Eastertown (Somerset)
Eastfield (North Yorkshire)
EASTGATE (Durham)
Eastgate (Lincolnshire)
Eastham (Merseyside)
Eastham Ferry (Merseyside)
Easthampton (Hereford & Worcester)
Easthope (Shropshire)
Easthorpe (Nottinghamshire)
Eastington (Gloucestershire)
Eastleach Martin (Gloucestershire)
Eastleach Turville (Gloucestershire)
Eastling (Kent)
Eastnor (Hereford & Worcester)
Eastoft (Lincolnshire)
EASTON (Bristol)
Easton (Cumbria)
Easton (Lincolnshire)
Easton (Somerset)
Easton-in-Gordano (Somerset)
Eastrington (East Riding of Yorkshire)
EASTRY (Kent)
EASTVILLE (Bristol)
Eastville (Lincolnshire)

See paras. 2 and 4 of the User Guide 2003 if you can't find your place name.

37

EASTWELL (Leicestershire)
EASTWOOD (Nottinghamshire)
Eastwood (West Yorkshire)
Eathorpe (Warwickshire)
Eaton (Cheshire)
EATON (Leicestershire)
EATON (Nottinghamshire)
Eaton (Shropshire)
Eaton Bishop (Hereford & Worcester)
Eaton Constantine (Shropshire)
Eaton Mascott (Shropshire)
Eaton upon Tern (Shropshire)
EAVES BROW (Cheshire)
EAVES GREEN (West Midlands)
Ebberston (North Yorkshire)
EBBW VALE (Blaenau Gwent)
EBCHESTER (Durham)
Ebdon (Somerset)
Ebley (Gloucestershire)
Ebnal (Cheshire)
Ebnall (Hereford & Worcester)
Ebrington (Gloucestershire)
ECCLES (Greater Manchester)
Eccles (Kent)
Eccles Green (Hereford & Worcester)
ECCLESALL (South Yorkshire)
ECCLESFIELD (South Yorkshire)
Eccleshall (Staffordshire)
ECCLESHILL (West Yorkshire)
Eccleston (Cheshire)
ECCLESTON (Lancashire)
ECCLESTON (Merseyside)
ECCLESTON GREEN (Lancashire)
ECKINGTON (Derbyshire)
Eckington (Hereford & Worcester)
Ecton (Staffordshire)
Edale (Derbyshire)
EDDERSIDE (Cumbria)
Eddington (Kent)
Eden Mount (Cumbria)
Edenbridge (Kent)
EDENFIELD (Lancashire)
Edenhall (Cumbria)
Edenham (Lincolnshire)
Edensor (Derbyshire)
EDENTHORPE (South Yorkshire)
Edgarley (Somerset)
EDGBASTON (West Midlands)
Edge (Gloucestershire)
EDGE (Shropshire)
EDGE END (Gloucestershire)
Edge Green (Cheshire)
Edge Hill (Merseyside)
Edgebolton (Shropshire)
EDGEFOLD (Greater Manchester)
Edgehill (Warwickshire)
Edgerley (Shropshire)
EDGERTON (West Yorkshire)
EDGERTOWN (West Yorkshire)
EDGESIDE (Lancashire)
Edgeworth (Gloucestershire)
Edgiock (Hereford & Worcester)
Edgmond (Shropshire)
Edgmond Marsh (Shropshire)
Edgton (Shropshire)
EDGWORTH (Lancashire)
Edial (Staffordshire)
EDINGALE (Staffordshire)
EDINGLEY (Nottinghamshire)
EDINGTON (Northumberland)
Edington (Somerset)
Edington Burtle (Somerset)
Edingworth (Somerset)
Edithmead (Somerset)
Edlingham (Northumberland)
Edlington (Lincolnshire)
Edmond Castle (Cumbria)
EDMONDSLEY (Durham)
Edmondthorpe (Leicestershire)
Edmundbyers (Durham)
Ednaston (Derbyshire)
Edstaston (Shropshire)
Edstone (Warwickshire)
Edvin Loach (Hereford & Worcester)
EDWALTON (Nottinghamshire)
EDWARDSVILLE (Merthyr Tydfil)
Edwinsford (Carmarthenshire)
EDWINSTOWE (Nottinghamshire)
Edwyn Ralph (Hereford & Worcester)
EFAIL ISAF (Rhondda Cynon Taff)
EFAIL-FACH (Neath Port Talbot)

Efail-rhyd (Powys)
Efailwen (Carmarthenshire)
Efenechtyd (Denbighshire)
Efflinch (Staffordshire)
EGERTON (Greater Manchester)
Egerton (Kent)
Egginton (Derbyshire)
Egglescliffe (Durham)
EGGLESTON (Durham)
EGLINGHAM (Northumberland)
EGLWYS CROSS (Wrexham)
Eglwys-Brewis (Vale of Glamorgan)
Eglwyswrw (Pembrokeshire)
EGMANTON (Nottinghamshire)
Egremont (Cumbria)
Egremont (Merseyside)
Egton (North Yorkshire)
Egton Bridge (North Yorkshire)
Eight and Forty (East Riding of Yorkshire)
Elan Village (Powys)
Elberton (Gloucestershire)
Eldersfield (Hereford & Worcester)
Eldmire (North Yorkshire)
ELDON (Durham)
ELDWICK (West Yorkshire)
ELFORD (Northumberland)
Elford (Staffordshire)
Elham (Kent)
Elilaw (Northumberland)
Elim (Anglesey)
ELISHAW (Northumberland)
ELKESLEY (Nottinghamshire)
Elkstone (Gloucestershire)
ELLAND (West Yorkshire)
ELLAND LOWER EDGE (West Yorkshire)
Ellastone (Staffordshire)
Ellel (Lancashire)
ELLENBOROUGH (Cumbria)
ELLENBROOK (Greater Manchester)
Ellenhall (Staffordshire)
Ellerbeck (North Yorkshire)
ELLERBY (North Yorkshire)
Ellerdine Heath (Shropshire)
Ellerker (East Riding of Yorkshire)
ELLERS (North Yorkshire)
ELLERTON (East Riding of Yorkshire)
Ellerton (North Yorkshire)
Ellerton (Shropshire)
Ellesmere (Shropshire)
Ellesmere Port (Cheshire)
Ellicombe (Somerset)
Ellingham (Northumberland)
Ellingstring (North Yorkshire)
ELLINGTON (Northumberland)
Elliots Green (Somerset)
ELLISTOWN (Leicestershire)
Ellonby (Cumbria)
Elloughton (East Riding of Yorkshire)
ELLWOOD (Gloucestershire)
Elmbridge (Hereford & Worcester)
Elmdon (West Midlands)
Elmdon Heath (West Midlands)
ELMER'S GREEN (Lancashire)
Elmesthorpe (Leicestershire)
Elmhurst (Staffordshire)
Elmley Castle (Hereford & Worcester)
Elmley Lovett (Hereford & Worcester)
Elmore (Gloucestershire)
Elmore Back (Gloucestershire)
ELMS GREEN (Hereford & Worcester)
Elmsted Court (Kent)
ELMSTONE (Kent)
Elmstone Hardwicke (Gloucestershire)
Elmswell (East Riding of Yorkshire)
ELMTON (Derbyshire)
Elrington (Northumberland)
ELSDON (Northumberland)
ELSECAR (South Yorkshire)
Elsham (Lincolnshire)
Elslack (North Yorkshire)
Elson (Shropshire)
Elsthorpe (Lincolnshire)
Elstob (Durham)
Elston (Lancashire)
Elston (Nottinghamshire)
Elstronwick (East Riding of Yorkshire)
Elswick (Lancashire)
ELSWICK (Tyne & Wear)
Elterwater (Cumbria)
Elton (Cheshire)
Elton (Derbyshire)

Elton (Durham)
Elton (Gloucestershire)
ELTON (Greater Manchester)
Elton (Hereford & Worcester)
Elton (Nottinghamshire)
Elton Green (Cheshire)
ELTRINGHAM (Northumberland)
Elvaston (Derbyshire)
ELVINGTON (Kent)
ELVINGTON (North Yorkshire)
Elwick (Durham)
Elwick (Northumberland)
Elworth (Cheshire)
Elworthy (Somerset)
Ely (Cardiff)
Embleton (Cumbria)
EMBLETON (Durham)
EMBLETON (Northumberland)
EMBOROUGH (Somerset)
EMBSAY (North Yorkshire)
EMLEY (West Yorkshire)
EMLEY MOOR (West Yorkshire)
EMMETT CARR (Derbyshire)
ENCHMARSH (Shropshire)
Enderby (Leicestershire)
Endmoor (Cumbria)
Endon (Staffordshire)
Endon Bank (Staffordshire)
ENGINE COMMON (Gloucestershire)
England's Gate (Hereford & Worcester)
Englesea-brook (Cheshire)
ENGLISH BICKNOR (Gloucestershire)
English Frankton (Shropshire)
Englishcombe (Somerset)
Enmore (Somerset)
Ennerdale Bridge (Cumbria)
Ensdon (Shropshire)
Enson (Staffordshire)
Enterpen (North Yorkshire)
Enville (Staffordshire)
Epney (Gloucestershire)
EPPERSTONE (Nottinghamshire)
EPPLEBY (North Yorkshire)
Eppleworth (East Riding of Yorkshire)
Epworth (Lincolnshire)
Epworth Turbary (Lincolnshire)
ERBISTOCK (Wrexham)
ERDINGTON (West Midlands)
Erriottwood (Kent)
Erwood (Powys)
Eryholme (North Yorkshire)
Eryrys (Denbighshire)
ESCOMB (Durham)
Escott (Somerset)
ESCRICK (North Yorkshire)
Esgair (Carmarthenshire)
Esgairgeiliog (Powys)
Esgerdawe (Carmarthenshire)
ESH (Durham)
ESH WINNING (Durham)
ESHOLT (West Yorkshire)
ESHOTT (Northumberland)
Eshton (North Yorkshire)
Eskdale Green (Cumbria)
ESKETT (Cumbria)
Eskham (Lincolnshire)
ESKHOLME (South Yorkshire)
ESPERLEY LANE ENDS (Durham)
Esprick (Lancashire)
ESSINGTON (Staffordshire)
Eston (North Yorkshire)
ETAL (Northumberland)
Etchinghill (Kent)
ETCHINGHILL (Staffordshire)
Etherdwick (East Riding of Yorkshire)
Etloe (Gloucestershire)
ETRURIA (Staffordshire)
Ettersgill (Durham)
Ettiley Heath (Cheshire)
ETTINGSHALL (West Midlands)
Ettington (Warwickshire)
Etton (East Riding of Yorkshire)
Etwall (Derbyshire)
EUDON GEORGE (Shropshire)
EUXTON (Lancashire)
Evancoyd (Powys)
Evedon (Lincolnshire)
Evelith (Shropshire)
Evenjobb (Powys)
Evenlode (Gloucestershire)
EVENWOOD (Durham)

See paras. 2 and 4 of the User Guide 2003 if you can't find your place name.

EVENWOOD GATE (Durham)
Evercreech (Somerset)
Everingham (East Riding of Yorkshire)
Everley (North Yorkshire)
Everthorpe (East Riding of Yorkshire)
Everton (Merseyside)
EVERTON (Nottinghamshire)
Evesbatch (Hereford & Worcester)
Evesham (Hereford & Worcester)
Evington (Leicestershire)
EWDEN VILLAGE (South Yorkshire)
Ewdness (Shropshire)
Ewell Minnis (Kent)
Ewen (Gloucestershire)
Ewenny (Vale of Glamorgan)
Ewerby (Lincolnshire)
Ewerby Thorpe (Lincolnshire)
EWESLEY (Northumberland)
EWLOE (Flintshire)
EWLOE GREEN (Flintshire)
EWOOD (Lancashire)
EWOOD BRIDGE (Lancashire)
Ewyas Harold (Hereford & Worcester)
Exebridge (Somerset)
Exelby (North Yorkshire)
Exford (Somerset)
EXFORDSGREEN (Shropshire)
EXHALL (Warwickshire)
EXLEY HEAD (West Yorkshire)
Exted (Kent)
Exton (Somerset)
Eyam (Derbyshire)
Eye (Hereford & Worcester)
Eye Kettleby (Leicestershire)
Eyhorne Street (Kent)
Eynsford (Kent)
EYTHORNE (Kent)
Eyton (Hereford & Worcester)
Eyton (Shropshire)
EYTON (Wrexham)
Eyton on Severn (Shropshire)
Eyton upon the Weald Moor (Shropshire)

F

Faceby (North Yorkshire)
Fachwen (Powys)
FACIT (Lancashire)
FACKLEY (Nottinghamshire)
Faddiley (Cheshire)
Fadmoor (North Yorkshire)
FAERDRE (Swansea)
FAGWYR (Swansea)
Failand (Somerset)
FAILSWORTH (Greater Manchester)
FAIRBURN (North Yorkshire)
FAIRFIELD (Derbyshire)
Fairfield (Hereford & Worcester)
Fairfield (Kent)
Fairford (Gloucestershire)
Fairford Park (Gloucestershire)
Fairhaven (Lancashire)
Fairoak (Staffordshire)
Fairseat (Kent)
Fairwater (Cardiff)
Faldingworth (Lincolnshire)
Falfield (Gloucestershire)
Fallgate (Derbyshire)
FALLODEN (Northumberland)
FALLOWFIELD (Greater Manchester)
FALLOWFIELD (Northumberland)
Falsgrave (North Yorkshire)
FALSTONE (Northumberland)
Fangdale Beck (North Yorkshire)
Fangfoss (East Riding of Yorkshire)
Far End (Cumbria)
FAR FOREST (Hereford & Worcester)
Far Green (Gloucestershire)
FAR MOOR (Greater Manchester)
Far Oakridge (Gloucestershire)
Far Sawrey (Cumbria)
Far Thorpe (Lincolnshire)
FARDEN (Shropshire)
FAREWELL (Staffordshire)
Farforth (Lincolnshire)
FARINGTON (Lancashire)
FARLAM (Cumbria)
FARLEIGH (Somerset)
Farleigh Hungerford (Somerset)
Farlesthorpe (Lincolnshire)
Farleton (Cumbria)

FARLETON (Lancashire)
Farley (Derbyshire)
Farley (Staffordshire)
Farleys End (Gloucestershire)
Farlington (North Yorkshire)
FARLOW (Shropshire)
FARM TOWN (Leicestershire)
FARMBOROUGH (Somerset)
Farmcote (Gloucestershire)
Farmcote (Shropshire)
Farmers (Carmarthenshire)
Farmington (Gloucestershire)
Farnah Green (Derbyshire)
Farnborough (Warwickshire)
Farndon (Cheshire)
Farndon (Nottinghamshire)
Farnham (North Yorkshire)
Farningham (Kent)
FARNLEY (North Yorkshire)
FARNLEY (West Yorkshire)
FARNLEY TYAS (West Yorkshire)
FARNSFIELD (Nottinghamshire)
FARNWORTH (Cheshire)
FARNWORTH (Greater Manchester)
FARRINGTON GURNEY (Somerset)
FARSLEY (West Yorkshire)
Farthing Green (Kent)
Farthingloe (Kent)
FARTOWN (West Yorkshire)
FATFIELD (Tyne & Wear)
Faugh (Cumbria)
Fauld (Staffordshire)
FAULKLAND (Somerset)
Fauls (Shropshire)
Faversham (Kent)
Fawdington (North Yorkshire)
Fawdon (Northumberland)
Fawfieldhead (Staffordshire)
Fawkham Green (Kent)
Fawley Chapel (Hereford & Worcester)
FAWNOG (Flintshire)
Faxfleet (East Riding of Yorkshire)
Fazakerley (Merseyside)
FAZELEY (Staffordshire)
Fearby (North Yorkshire)
FEARNHEAD (Cheshire)
FEATHERSTONE (Staffordshire)
FEATHERSTONE (West Yorkshire)
Feckenham (Hereford & Worcester)
Feetham (North Yorkshire)
Feizor (North Yorkshire)
Felin gwm Isaf (Carmarthenshire)
Felin gwm Uchaf (Carmarthenshire)
Felin-newydd (Powys)
Felindre (Carmarthenshire)
Felindre (Powys)
Felindre (Swansea)
Felindre Farchog (Pembrokeshire)
Felinfach (Powys)
Felinfoel (Carmarthenshire)
Felixkirk (North Yorkshire)
FELKINGTON (Northumberland)
FELKIRK (West Yorkshire)
Fell Foot (Cumbria)
FELL LANE (West Yorkshire)
Fell Side (Cumbria)
FELLING (Tyne & Wear)
Felton (Hereford & Worcester)
Felton (Northumberland)
Felton (Somerset)
Felton Butler (Shropshire)
Fen End (Lincolnshire)
Fen End (West Midlands)
FENAY BRIDGE (West Yorkshire)
FENCE (Lancashire)
FENCE (South Yorkshire)
FENCEHOUSES (Tyne & Wear)
Fencote (North Yorkshire)
Fendike Corner (Lincolnshire)
Fenham (Northumberland)
FENHAM (Tyne & Wear)
FENISCLIFFE (Lancashire)
FENISCOWLES (Lancashire)
FENN GREEN (Shropshire)
Fenn Street (Kent)
Fenny Bentley (Derbyshire)
Fenny Compton (Warwickshire)
Fenny Drayton (Leicestershire)
FENROTHER (Northumberland)
Fenton (Cumbria)
Fenton (Lincolnshire)

Fenton (Northumberland)
Fenton (Nottinghamshire)
FENTON (Staffordshire)
FENWICK (Northumberland)
FENWICK (South Yorkshire)
FERNDALE (Rhondda Cynon Taff)
Fernhill Heath (Hereford & Worcester)
FERNILEE (Derbyshire)
Ferny Common (Hereford & Worcester)
Ferrensby (North Yorkshire)
Ferriby Sluice (Lincolnshire)
FERRYBRIDGE (West Yorkshire)
FERRYHILL (Durham)
Ferryside (Carmarthenshire)
Fewston (North Yorkshire)
Ffairfach (Carmarthenshire)
Ffald-y-Brenin (Carmarthenshire)
Ffawyddog (Powys)
Ffordd-Las (Denbighshire)
FFOREST (Carmarthenshire)
Fforest (Monmouthshire)
FFOREST FACH (Swansea)
FFOREST GOCH (Neath Port Talbot)
FFRITH (Flintshire)
FFYNNONGROEW (Flintshire)
Fiddington (Gloucestershire)
Fiddington (Somerset)
Field (Staffordshire)
Field Broughton (Cumbria)
Field Head (Leicestershire)
Fieldhead (Cumbria)
Filey (North Yorkshire)
Fillingham (Lincolnshire)
FILLONGLEY (Warwickshire)
FILTON (Bristol)
FILTON (Gloucestershire)
Fimber (East Riding of Yorkshire)
Findern (Derbyshire)
Finghall (North Yorkshire)
Fingland (Cumbria)
FINGLESHAM (Kent)
FINKLE STREET (South Yorkshire)
FINNINGLEY (South Yorkshire)
Finstall (Hereford & Worcester)
Finsthwaite (Cumbria)
FIR TREE (Durham)
Firbank (Cumbria)
FIRBECK (South Yorkshire)
Firby (North Yorkshire)
FIRGROVE (Greater Manchester)
Firsby (Lincolnshire)
FISHBURN (Durham)
Fisher's Row (Lancashire)
Fisherwick (Staffordshire)
Fishguard (Pembrokeshire)
Fishinghurst (Kent)
FISHLAKE (South Yorkshire)
Fishmere End (Lincolnshire)
FISHPONDS (Bristol)
FISHPOOL (Greater Manchester)
Fishtoft (Lincolnshire)
Fishtoft Drove (Lincolnshire)
Fishwick (Lancashire)
Fiskerton (Lincolnshire)
Fiskerton (Nottinghamshire)
Fitling (East Riding of Yorkshire)
Fitz (Shropshire)
Fitzhead (Somerset)
Fitzroy (Somerset)
FITZWILLIAM (West Yorkshire)
Five Bells (Somerset)
Five Bridges (Hereford & Worcester)
Five Lanes (Monmouthshire)
Five Oak Green (Kent)
FIVE ROADS (Carmarthenshire)
FIVE WAYS (Warwickshire)
Five Wents (Kent)
Fivecrosses (Cheshire)
Fivehead (Somerset)
Fladbury (Hereford & Worcester)
Flagg (Derbyshire)
Flamborough (East Riding of Yorkshire)
FLANSHAW (West Yorkshire)
FLAPPIT SPRING (West Yorkshire)
Flasby (North Yorkshire)
FLASH (Staffordshire)
Flawborough (Nottinghamshire)
Flawith (North Yorkshire)
Flax Bourton (Somerset)
Flaxby (North Yorkshire)
Flaxley (Gloucestershire)

See paras. 2 and 4 of the User Guide 2003 if you can't find your place name.

Flaxmere (Cheshire)
Flaxpool (Somerset)
Flaxton (North Yorkshire)
Fleckney (Leicestershire)
Flecknoe (Warwickshire)
Fledborough (Nottinghamshire)
Fleet (Lincolnshire)
Fleet Hargate (Lincolnshire)
Fleetwood (Lancashire)
Flemingston (Vale of Glamorgan)
Fletcher Green (Kent)
FLETCHERTOWN (Cumbria)
FLIMBY (Cumbria)
FLINT (Flintshire)
FLINT MOUNTAIN (Flintshire)
FLINT'S GREEN (West Midlands)
Flintham (Nottinghamshire)
Flinton (East Riding of Yorkshire)
Flixborough (Lincolnshire)
Flixborough Stather (Lincolnshire)
Flixton (Greater Manchester)
Flixton (North Yorkshire)
FLOCKTON (West Yorkshire)
FLOCKTON GREEN (West Yorkshire)
Flodden (Northumberland)
Flookburgh (Cumbria)
FLOTTERTON (Northumberland)
FLUSHDYKE (West Yorkshire)
Flyford Flavell (Hereford & Worcester)
FOCHRIW (Caerphilly)
Fockerby (Lincolnshire)
Foddington (Somerset)
Foel (Powys)
FOEL Y DYFFRYN (Bridgend)
FOELGASTELL (Carmarthenshire)
Foggathorpe (East Riding of Yorkshire)
Fole (Staffordshire)
FOLESHILL (West Midlands)
Folkestone (Kent)
Folkingham (Lincolnshire)
Folkton (North Yorkshire)
Follifoot (North Yorkshire)
Fonmon (Vale of Glamorgan)
Font-y-gary (Vale of Glamorgan)
Foolow (Derbyshire)
FORCETT (North Yorkshire)
FORD (Derbyshire)
Ford (Gloucestershire)
FORD (Northumberland)
Ford (Shropshire)
Ford (Somerset)
Ford (Staffordshire)
Ford Green (Lancashire)
FORD HEATH (Shropshire)
Ford Street (Somerset)
Fordcombe (Kent)
Forden (Powys)
Fordon (East Riding of Yorkshire)
Fordwich (Kent)
Forebridge (Staffordshire)
Foremark (Derbyshire)
Forest (North Yorkshire)
Forest Becks (Lancashire)
Forest Hall (Cumbria)
Forest Hall (Tyne & Wear)
FOREST HEAD (Cumbria)
Forest Lane Head (North Yorkshire)
FOREST TOWN (Nottinghamshire)
Forest-in-Teesdale (Durham)
FORESTBURN GATE (Northumberland)
Forge (Powys)
Forge Hammer (Torfaen)
FORGE SIDE (Torfaen)
Forhill (Hereford & Worcester)
Formby (Merseyside)
Fornside (Cumbria)
FORSBROOK (Staffordshire)
Forshaw Heath (Warwickshire)
Forthampton (Gloucestershire)
Forton (Lancashire)
Forton (Shropshire)
Forton (Somerset)
Forton (Staffordshire)
Fosdyke (Lincolnshire)
Fosdyke Bridge (Lincolnshire)
Fossebridge (Gloucestershire)
FOSTERHOUSES (South Yorkshire)
Foston (Derbyshire)
Foston (Leicestershire)
Foston (Lincolnshire)
Foston (North Yorkshire)

Foston on the Wolds (East Riding of Yorkshire)
Fotherby (Lincolnshire)
FOTHERGILL (Cumbria)
FOUL END (Warwickshire)
Foulbridge (Cumbria)
FOULBY (West Yorkshire)
Foulridge (Lancashire)
FOUR ASHES (Staffordshire)
Four Ashes (West Midlands)
Four Crosses (Powys)
FOUR CROSSES (Staffordshire)
Four Elms (Kent)
Four Foot (Somerset)
Four Forks (Somerset)
FOUR GATES (Greater Manchester)
FOUR LANE END (South Yorkshire)
Four Lane Ends (Cheshire)
Four Mile Bridge (Anglesey)
FOUR OAKS (Gloucestershire)
FOUR OAKS (West Midlands)
Four Roads (Carmarthenshire)
Four Shire Stone (Warwickshire)
Four Throws (Kent)
Four Wents (Kent)
Fourlanes End (Cheshire)
FOURSTONES (Northumberland)
FOWLEY COMMON (Cheshire)
Fowlhall (Kent)
Fownhope (Hereford & Worcester)
Foxcote (Gloucestershire)
FOXCOTE (Somerset)
Foxendown (Kent)
Foxfield (Cumbria)
FOXHOLE (Swansea)
Foxholes (North Yorkshire)
Foxlydiate (Hereford & Worcester)
FOXT (Staffordshire)
Foxton (Durham)
Foxton (Leicestershire)
Foxton (North Yorkshire)
Foxup (North Yorkshire)
Foxwist Green (Cheshire)
FOXWOOD (Shropshire)
Foy (Hereford & Worcester)
Fradley (Staffordshire)
Fradswell (Staffordshire)
Fraisthorpe (East Riding of Yorkshire)
Frampton (Lincolnshire)
FRAMPTON COTTERELL (Gloucestershire)
Frampton Mansell (Gloucestershire)
Frampton on Severn (Gloucestershire)
Frampton West End (Lincolnshire)
FRAMWELLGATE MOOR (Durham)
Frances Green (Lancashire)
Franche (Hereford & Worcester)
Frandley (Cheshire)
Frank's Bridge (Powys)
Frankby (Merseyside)
Franklands Gate (Hereford & Worcester)
Frankley (Hereford & Worcester)
Frankton (Warwickshire)
Freckleton (Lancashire)
FREEBIRCH (Derbyshire)
Freeby (Leicestershire)
FREEHAY (Staffordshire)
Freiston (Lincolnshire)
FREMINGTON (North Yorkshire)
French Street (Kent)
FRENCHAY (Gloucestershire)
Freshfield (Merseyside)
Freshwater East (Pembrokeshire)
Fretherne (Gloucestershire)
FREYSTROP (Pembrokeshire)
Friars' Hill (North Yorkshire)
Fridaythorpe (East Riding of Yorkshire)
Friden (Derbyshire)
FRIENDLY (West Yorkshire)
Friesthorpe (Lincolnshire)
Frieston (Lincolnshire)
FRIEZELAND (Nottinghamshire)
Frindsbury (Kent)
Frinsted (Kent)
Frisby on the Wreake (Leicestershire)
Friskney (Lincolnshire)
Friskney Eaudike (Lincolnshire)
FRITCHLEY (Derbyshire)
Frith Bank (Lincolnshire)
FRITH COMMON (Hereford & Worcester)
Frithville (Lincolnshire)
Frittenden (Kent)
FRIZINGHALL (West Yorkshire)

FRIZINGTON (Cumbria)
Frocester (Gloucestershire)
FRODESLEY (Shropshire)
Frodsham (Cheshire)
Frog Pool (Hereford & Worcester)
Froggatt (Derbyshire)
FROGHALL (Staffordshire)
FROGHAM (Kent)
Frognall (Lincolnshire)
Frolesworth (Leicestershire)
Frome (Somerset)
Fromes Hill (Hereford & Worcester)
Fron (Powys)
Fron Isaf (Wrexham)
Froncysyllte (Denbighshire)
Frosterley (Durham)
Fryton (North Yorkshire)
Fulbeck (Lincolnshire)
Fulford (North Yorkshire)
Fulford (Somerset)
Fulford (Staffordshire)
Full Sutton (East Riding of Yorkshire)
Fuller Street (Kent)
Fuller's Moor (Cheshire)
Fulletby (Lincolnshire)
Fullready (Warwickshire)
FULNECK (West Yorkshire)
Fulnetby (Lincolnshire)
Fulney (Lincolnshire)
FULSTONE (West Yorkshire)
Fulstow (Lincolnshire)
Fulwood (Lancashire)
FULWOOD (Nottinghamshire)
Fulwood (Somerset)
FULWOOD (South Yorkshire)
FURNACE (Carmarthenshire)
FURNACE END (Warwickshire)
FURNESS VALE (Derbyshire)
Further Quarter (Kent)
Furzehills (Lincolnshire)
Fyfett (Somerset)
FYLINGTHORPE (North Yorkshire)

G

Gaddesby (Leicestershire)
Gadfa (Anglesey)
Gadlas (Shropshire)
Gaer (Powys)
Gaer-llwyd (Monmouthshire)
Gaerwen (Anglesey)
Gailey (Staffordshire)
GAINFORD (Durham)
Gainsborough (Lincolnshire)
GAISBY (West Yorkshire)
Gaisgill (Cumbria)
GAITSGILL (Cumbria)
Galby (Leicestershire)
Galgate (Lancashire)
Galhampton (Somerset)
Gallantry Bank (Cheshire)
GALLEY COMMON (Warwickshire)
Gallows Green (Hereford & Worcester)
Galphay (North Yorkshire)
GAMBALLS GREEN (Staffordshire)
Gamblesby (Cumbria)
Gamelsby (Cumbria)
GAMESLEY (Greater Manchester)
GAMMERSGILL (North Yorkshire)
GAMSTON (Nottinghamshire)
Ganarew (Hereford & Worcester)
Ganstead (East Riding of Yorkshire)
Ganthorpe (North Yorkshire)
Ganton (North Yorkshire)
GARDEN CITY (Flintshire)
GARDEN VILLAGE (Derbyshire)
Gardham (East Riding of Yorkshire)
Gare Hill (Somerset)
GARFORTH (West Yorkshire)
GARFORTH BRIDGE (West Yorkshire)
Gargrave (North Yorkshire)
Garlinge (Kent)
Garlinge Green (Kent)
GARMONDSWAY (Durham)
GARMSTON (Shropshire)
GARNANT (Carmarthenshire)
Garnett Bridge (Cumbria)
GARNSWLLT (Swansea)
GARRIGILL (Cumbria)
Garriston (North Yorkshire)
Garrowby Hall (East Riding of Yorkshire)

See paras. 2 and 4 of the User Guide 2003 if you can't find your place name.

Garsdale (Cumbria)
Garsdale Head (Cumbria)
Garshall Green (Staffordshire)
Garstang (Lancashire)
Garston (Merseyside)
Garth (Denbighshire)
Garth (Monmouthshire)
Garth (Powys)
Garth Row (Cumbria)
Garthbrengy (Powys)
Garthmyl (Powys)
Garthorpe (Leicestershire)
Garthorpe (Lincolnshire)
Garths (Cumbria)
Garton (East Riding of Yorkshire)
Garton-on-the-Wolds (East Riding of Yorkshire)
Garway (Hereford & Worcester)
Garway Common (Hereford & Worcester)
Garway Hill (Hereford & Worcester)
Gate Burton (Lincolnshire)
Gate Helmsley (North Yorkshire)
Gateacre (Merseyside)
Gatebeck (Cumbria)
GATEFORD (Nottinghamshire)
GATEFORTH (North Yorkshire)
GATEHOUSE (Northumberland)
Gatenby (North Yorkshire)
Gates Heath (Cheshire)
Gatesgarth (Cumbria)
GATESHEAD (Tyne & Wear)
GATHURST (Greater Manchester)
Gatley (Greater Manchester)
Gaufron (Powys)
GAULKTHORN (Lancashire)
Gaunton's Bank (Cheshire)
Gautby (Lincolnshire)
GAWBER (South Yorkshire)
Gawsworth (Cheshire)
GAWTHORPE (West Yorkshire)
Gawthrop (Cumbria)
Gawthwaite (Cumbria)
Gaydon (Warwickshire)
GAYLE (North Yorkshire)
GAYLES (North Yorkshire)
Gayton (Merseyside)
Gayton (Staffordshire)
Gayton le Marsh (Lincolnshire)
Geddinge (Kent)
GEDLING (Nottinghamshire)
Gedney (Lincolnshire)
Gedney Broadgate (Lincolnshire)
Gedney Drove End (Lincolnshire)
Gedney Dyke (Lincolnshire)
Gedney Hill (Lincolnshire)
GEE CROSS (Greater Manchester)
GELLI (Rhondda Cynon Taff)
Gelli Gynan (Denbighshire)
Gellifor (Denbighshire)
GELLIGAER (Caerphilly)
GELLIGROES (Caerphilly)
GELLIGRON (Neath Port Talbot)
GELLINUDD (Neath Port Talbot)
Gelly (Pembrokeshire)
Gellywen (Carmarthenshire)
Gelston (Lincolnshire)
Gembling (East Riding of Yorkshire)
GENTLESHAW (Staffordshire)
GERRICK (North Yorkshire)
Geuffordd (Powys)
Gib Hill (Cheshire)
Gibraltar (Lincolnshire)
Gibsmere (Nottinghamshire)
Giggleswick (North Yorkshire)
Gilberdyke (East Riding of Yorkshire)
Gilbert's Cross (Staffordshire)
Gilbert's End (Hereford & Worcester)
GILCRUX (Cumbria)
GILDERSOME (West Yorkshire)
GILDINGWELLS (South Yorkshire)
GILESGATE MOOR (Durham)
Gileston (Vale of Glamorgan)
GILFACH (Caerphilly)
GILFACH GOCH (Bridgend)
GILGARRAN (Cumbria)
Gill (Cumbria)
Gill's Green (Kent)
GILLAMOOR (North Yorkshire)
GILLING EAST (North Yorkshire)
Gilling West (North Yorkshire)
Gillingham (Kent)
Gillmoss (Merseyside)

GILLOW HEATH (Staffordshire)
Gilmonby (Durham)
Gilmorton (Leicestershire)
Gilsland (Northumberland)
Gilson (Warwickshire)
GILSTEAD (West Yorkshire)
GILTBROOK (Nottinghamshire)
Gilwern (Monmouthshire)
GINCLOUGH (Cheshire)
Gipsey Bridge (Lincolnshire)
GIRLINGTON (West Yorkshire)
Girsby (North Yorkshire)
GIRTON (Nottinghamshire)
Gisburn (Lancashire)
Gladestry (Powys)
GLAIS (Swansea)
GLAISDALE (North Yorkshire)
Glan-Duar (Carmarthenshire)
Glan-Mule (Powys)
GLAN-RHYD (Powys)
GLAN-Y-DON (Flintshire)
GLAN-Y-LLYN (Rhondda Cynon Taff)
Glan-y-nant (Powys)
Glan-yr-afon (Anglesey)
Glanafon (Pembrokeshire)
GLANAMAN (Carmarthenshire)
Glandwr (Pembrokeshire)
Glangrwyne (Powys)
Glanrhyd (Pembrokeshire)
Glanton (Northumberland)
Glanton Pike (Northumberland)
GLAPWELL (Derbyshire)
Glasbury (Powys)
Glascoed (Denbighshire)
Glascoed (Monmouthshire)
GLASCOTE (Staffordshire)
Glascwm (Powys)
Glaspwll (Powys)
GLASS HOUGHTON (West Yorkshire)
Glassenbury (Kent)
Glasshouse (Gloucestershire)
Glasshouse Hill (Gloucestershire)
GLASSHOUSES (North Yorkshire)
Glasson (Cumbria)
Glasson (Lancashire)
Glassonby (Cumbria)
Glastonbury (Somerset)
GLAZEBROOK (Cheshire)
GLAZEBURY (Cheshire)
GLAZELEY (Shropshire)
Gleadsmoss (Cheshire)
Gleaston (Cumbria)
GLEDHOW (West Yorkshire)
GLEDRID (Shropshire)
Glen Parva (Leicestershire)
Glenfield (Leicestershire)
Glenridding (Cumbria)
Glentham (Lincolnshire)
Glentworth (Lincolnshire)
Glewstone (Hereford & Worcester)
Glooston (Leicestershire)
GLORORUM (Northumberland)
Glossop (Derbyshire)
GLOSTER HILL (Northumberland)
Gloucester (Gloucestershire)
GLUSBURN (North Yorkshire)
Glyn Ceiriog (Wrexham)
GLYN-NEATH (Neath Port Talbot)
GLYNCORRWG (Neath Port Talbot)
Glyndyfrdwy (Denbighshire)
GLYNTAFF (Rhondda Cynon Taff)
Glyntawe (Powys)
Glynteg (Carmarthenshire)
Gnosall (Staffordshire)
Gnosall Heath (Staffordshire)
Goadby (Leicestershire)
GOADBY MARWOOD (Leicestershire)
Goat Lees (Kent)
GOATHLAND (North Yorkshire)
Goathurst (Somerset)
Goathurst Common (Kent)
GOBOWEN (Shropshire)
Goddard's Green (Kent)
GODLEY (Greater Manchester)
Godmersham (Kent)
Godney (Somerset)
GODRE'R-GRAIG (Neath Port Talbot)
Godstone (Staffordshire)
Goetre (Monmouthshire)
Gofilon (Monmouthshire)
GOLBORNE (Greater Manchester)

GOLCAR (West Yorkshire)
Goldcliff (Newport)
Golden Green (Kent)
Golden Grove (Carmarthenshire)
Golden Hill (Pembrokeshire)
GOLDEN VALLEY (Derbyshire)
GOLDENHILL (Staffordshire)
Golding (Shropshire)
GOLDS GREEN (West Midlands)
GOLDSBOROUGH (North Yorkshire)
Goldstone (Kent)
Goldstone (Shropshire)
GOLDTHORPE (South Yorkshire)
Golford (Kent)
Golford Green (Kent)
GOLLINGLITH FOOT (North Yorkshire)
GOLLY (Wrexham)
Golsoncott (Somerset)
GOMERSAL (West Yorkshire)
GONALSTON (Nottinghamshire)
Gonerby Hill Foot (Lincolnshire)
Goodmanham (East Riding of Yorkshire)
GOODNESTONE (Kent)
Goodrich (Hereford & Worcester)
GOODSHAW (Lancashire)
GOODSHAW FOLD (Lancashire)
Goodwick (Pembrokeshire)
GOODYERS END (Warwickshire)
GOOLE (East Riding of Yorkshire)
Goom's Hill (Hereford & Worcester)
GOOSE GREEN (Gloucestershire)
GOOSE GREEN (Greater Manchester)
Goose Green (Kent)
Goose Pool (Hereford & Worcester)
Goosehill Green (Hereford & Worcester)
Goosemoor (Somerset)
Goosnargh (Lancashire)
Goostrey (Cheshire)
Gore (Powys)
Gore Street (Kent)
Gorsedd (Flintshire)
GORSEINON (Swansea)
Gorseybank (Derbyshire)
Gorslas (Carmarthenshire)
GORSLEY (Gloucestershire)
Gorsley Common (Gloucestershire)
GORST HILL (Hereford & Worcester)
Gorstage (Cheshire)
Gorstello (Cheshire)
Gorsty Common (Hereford & Worcester)
Gorsty Hill (Staffordshire)
GORTON (Greater Manchester)
Gosberton (Lincolnshire)
Gosberton Clough (Lincolnshire)
Gosforth (Cumbria)
GOSFORTH (Tyne & Wear)
Gosland Green (Cheshire)
Gosling Street (Somerset)
GOSPEL END (Staffordshire)
Gossington (Gloucestershire)
Goswick (Northumberland)
Gotham (Nottinghamshire)
Gotherington (Gloucestershire)
Gotton (Somerset)
Goudhurst (Kent)
Goulceby (Lincolnshire)
GOWDALL (East Riding of Yorkshire)
GOWERTON (Swansea)
Gowthorpe (East Riding of Yorkshire)
Goxhill (East Riding of Yorkshire)
Goxhill (Lincolnshire)
Graby (Lincolnshire)
Gradeley Green (Cheshire)
Grafton (Hereford & Worcester)
Grafton (North Yorkshire)
Grafton (Shropshire)
Grafton Flyford (Hereford & Worcester)
Grafty Green (Kent)
Graianrhyd (Denbighshire)
Graig (Denbighshire)
Graig-fechan (Denbighshire)
Grain (Kent)
GRAINS BAR (Greater Manchester)
Grainsby (Lincolnshire)
Grainthorpe (Lincolnshire)
Graiselound (Lincolnshire)
Granby (Merseyside)
GRANBY (Nottinghamshire)
Grandborough (Warwickshire)
Grange (Cumbria)
Grange (Kent)

See paras. 2 and 4 of the User Guide 2003 if you can't find your place name.

41

Grange (Merseyside)
GRANGE MOOR (West Yorkshire)
GRANGE VILLA (Durham)
Grange-over-Sands (Cumbria)
Grangemill (Derbyshire)
Grangetown (North Yorkshire)
Gransmoor (East Riding of Yorkshire)
Granston (Pembrokeshire)
Grantham (Lincolnshire)
Grantsfield (Hereford & Worcester)
Grappenhall (Cheshire)
Grasby (Lincolnshire)
Grasmere (Cumbria)
GRASSCROFT (Greater Manchester)
Grassendale (Merseyside)
GRASSGARTH (Cumbria)
GRASSINGTON (North Yorkshire)
GRASSMOOR (Derbyshire)
GRASSTHORPE (Nottinghamshire)
Gratwich (Staffordshire)
GRAVELLY HILL (West Midlands)
Gravelsbank (Shropshire)
Graveney (Kent)
Gravesend (Kent)
Grayingham (Lincolnshire)
Grayrigg (Cumbria)
GRAYSON GREEN (Cumbria)
Graythorpe (Durham)
GREASBROUGH (South Yorkshire)
Greasby (Merseyside)
GREASLEY (Nottinghamshire)
Great Alne (Warwickshire)
Great Altcar (Lancashire)
GREAT ASBY (Cumbria)
Great Ayton (North Yorkshire)
Great Badminton (Gloucestershire)
GREAT BARR (West Midlands)
Great Barrington (Gloucestershire)
Great Barrow (Cheshire)
Great Barugh (North Yorkshire)
Great Bavington (Northumberland)
Great Blencow (Cumbria)
Great Bolas (Shropshire)
Great Bowden (Leicestershire)
GREAT BRIDGE (West Midlands)
Great Bridgeford (Staffordshire)
GREAT BROUGHTON (Cumbria)
Great Broughton (North Yorkshire)
Great Budworth (Cheshire)
Great Burdon (Durham)
Great Busby (North Yorkshire)
Great Carlton (Lincolnshire)
Great Chart (Kent)
Great Chatwell (Staffordshire)
GREAT CHELL (Staffordshire)
GREAT CLIFFE (West Yorkshire)
GREAT CLIFTON (Cumbria)
Great Coates (Lincolnshire)
Great Comberton (Hereford & Worcester)
Great Comp (Kent)
Great Corby (Cumbria)
Great Cowden (East Riding of Yorkshire)
Great Crosby (Merseyside)
Great Crosthwaite (Cumbria)
Great Cubley (Derbyshire)
Great Dalby (Leicestershire)
Great Doward (Hereford & Worcester)
Great Easton (Leicestershire)
Great Eccleston (Lancashire)
Great Edstone (North Yorkshire)
GREAT ELM (Somerset)
Great Givendale (East Riding of Yorkshire)
Great Glen (Leicestershire)
Great Gonerby (Lincolnshire)
Great Habton (North Yorkshire)
Great Hale (Lincolnshire)
GREAT HANWOOD (Shropshire)
GREAT HARWOOD (Lancashire)
Great Hatfield (East Riding of Yorkshire)
GREAT HAYWOOD (Staffordshire)
GREAT HECK (North Yorkshire)
GREAT HORTON (West Yorkshire)
GREAT HOUGHTON (South Yorkshire)
Great Hucklow (Derbyshire)
Great Kelk (East Riding of Yorkshire)
Great Langdale (Cumbria)
Great Langton (North Yorkshire)
Great Limber (Lincolnshire)
Great Longstone (Derbyshire)
GREAT LUMLEY (Durham)
GREAT LYTH (Shropshire)

Great Malvern (Hereford & Worcester)
Great Marton (Lancashire)
Great Meols (Merseyside)
Great Mitton (Lancashire)
GREAT MONGEHAM (Kent)
Great Musgrave (Cumbria)
Great Ness (Shropshire)
Great Nurcott (Somerset)
Great Oak (Monmouthshire)
Great Ormside (Cumbria)
Great Orton (Cumbria)
Great Ouseburn (North Yorkshire)
Great Pattenden (Kent)
Great Plumpton (Lancashire)
Great Ponton (Lincolnshire)
GREAT PRESTON (West Yorkshire)
Great Rissington (Gloucestershire)
Great Rudbaxton (Pembrokeshire)
Great Ryle (Northumberland)
GREAT RYTON (Shropshire)
Great Salkeld (Cumbria)
GREAT SANKEY (Cheshire)
GREAT SAREDON (Staffordshire)
Great Saughall (Cheshire)
Great Smeaton (North Yorkshire)
Great Soudley (Shropshire)
Great Stainton (Durham)
Great Steeping (Lincolnshire)
Great Stonar (Kent)
GREAT STRICKLAND (Cumbria)
Great Sturton (Lincolnshire)
Great Sutton (Cheshire)
Great Sutton (Shropshire)
Great Swinburne (Northumberland)
Great Tosson (Northumberland)
Great Tows (Lincolnshire)
Great Urswick (Cumbria)
Great Washbourne (Gloucestershire)
GREAT WHITTINGTON (Northumberland)
Great Witcombe (Gloucestershire)
GREAT WITLEY (Hereford & Worcester)
Great Wolford (Warwickshire)
GREAT WYRLEY (Staffordshire)
Great Wytheford (Shropshire)
Greatford (Lincolnshire)
Greatgate (Staffordshire)
Greatham (Durham)
Greatstone-on-Sea (Kent)
Grebby (Lincolnshire)
Green (Denbighshire)
Green Bank (Cumbria)
Green Down (Somerset)
GREEN END (Warwickshire)
Green Hammerton (North Yorkshire)
Green Head (Cumbria)
GREEN HEATH (Staffordshire)
Green Lane (Hereford & Worcester)
GREEN MOOR (South Yorkshire)
Green Oak (East Riding of Yorkshire)
Green Ore (Somerset)
Green Quarter (Cumbria)
Green Street (Gloucestershire)
Green Street (Hereford & Worcester)
Green Street Green (Kent)
GREENCROFT HALL (Durham)
GREENFIELD (Flintshire)
Greenfield (Greater Manchester)
GREENGATES (West Yorkshire)
GREENGILL (Cumbria)
Greenhalgh (Lancashire)
Greenham (Somerset)
Greenhaugh (Northumberland)
Greenhead (Northumberland)
GREENHEYS (Greater Manchester)
Greenhill (Hereford & Worcester)
Greenhill (Kent)
GREENHILLOCKS (Derbyshire)
Greenhithe (Kent)
Greenholme (Cumbria)
GREENHOW HILL (North Yorkshire)
GREENLAND (South Yorkshire)
GREENMOUNT (Greater Manchester)
Greenodd (Cumbria)
GREENSIDE (Tyne & Wear)
GREENSIDE (West Yorkshire)
Greenway (Gloucestershire)
GREENWAY (Hereford & Worcester)
Greenway (Somerset)
Greenway (Vale of Glamorgan)
Greet (Gloucestershire)
Greete (Shropshire)

Greetham (Lincolnshire)
GREETLAND (West Yorkshire)
GREGSON LANE (Lancashire)
Greinton (Somerset)
GRENDON (Warwickshire)
Grendon Green (Hereford & Worcester)
GRENOSIDE (South Yorkshire)
GRESFORD (Wrexham)
GRESSINGHAM (Lancashire)
Gresty Green (Cheshire)
Greta Bridge (Durham)
Gretton (Gloucestershire)
GRETTON (Shropshire)
Grewelthorpe (North Yorkshire)
Grey Green (Lincolnshire)
GREYGARTH (North Yorkshire)
Greylake (Somerset)
GREYSOUTHEN (Cumbria)
Greystoke (Cumbria)
Gribthorpe (East Riding of Yorkshire)
GRIFF (Warwickshire)
Griffithstown (Torfaen)
GRIFFYDAM (Leicestershire)
GRIMEFORD VILLAGE (Lancashire)
GRIMESTHORPE (South Yorkshire)
GRIMETHORPE (South Yorkshire)
Grimley (Hereford & Worcester)
Grimoldby (Lincolnshire)
Grimpo (Shropshire)
Grimsargh (Lancashire)
Grimsby (Lincolnshire)
GRIMSHAW (Lancashire)
GRIMSHAW GREEN (Lancashire)
Grimsthorpe (Lincolnshire)
Grimston (East Riding of Yorkshire)
GRIMSTON (Leicestershire)
GRIMSTON HILL (Nottinghamshire)
Grindale (East Riding of Yorkshire)
GRINDLE (Shropshire)
Grindleford (Derbyshire)
Grindleton (Lancashire)
Grindley Brook (Shropshire)
Grindlow (Derbyshire)
Grindon (Durham)
Grindon (Northumberland)
Grindon (Staffordshire)
GRINDON HILL (Northumberland)
GRINDONRIGG (Northumberland)
GRINGLEY ON THE HILL (Nottinghamshire)
Grinsdale (Cumbria)
Grinshill (Shropshire)
Grinton (North Yorkshire)
Gristhorpe (North Yorkshire)
Grizebeck (Cumbria)
Grizedale (Cumbria)
Groby (Leicestershire)
GROES-FAEN (Rhondda Cynon Taff)
GROES-WEN (Caerphilly)
Groesffordd Marli (Denbighshire)
Groesllwyd (Powys)
Gronant (Flintshire)
Grosmont (Monmouthshire)
Grosmont (North Yorkshire)
GROTTON (Greater Manchester)
GROVE (Kent)
Grove (Nottinghamshire)
Grove (Pembrokeshire)
Grove Green (Kent)
Grove Vale (West Midlands)
Grovenhurst (Kent)
Grovesend (Gloucestershire)
GROVESEND (Swansea)
Grubb Street (Kent)
Guanockgate (Lincolnshire)
Guarlford (Hereford & Worcester)
GUIDE (Lancashire)
GUIDE BRIDGE (Greater Manchester)
GUIDE POST (Northumberland)
Guilden Down (Shropshire)
Guilden Sutton (Cheshire)
Guildstead (Kent)
Guilsfield (Powys)
Guilton (Kent)
Guisborough (North Yorkshire)
GUISELEY (West Yorkshire)
Guiting Power (Gloucestershire)
GUMFRESTON (Pembrokeshire)
Gumley (Leicestershire)
Gun Green (Kent)
GUN HILL (Warwickshire)
GUNBY (East Riding of Yorkshire)

See paras. 2 and 4 of the User Guide 2003 if you can't find your place name.

Gunby (Lincolnshire)
Gunnerside (North Yorkshire)
GUNNERTON (Northumberland)
Gunness (Lincolnshire)
GUNTHORPE (Nottinghamshire)
Gupworthy (Somerset)
GURNETT (Cheshire)
GURNEY SLADE (Somerset)
GURNOS (Powys)
Gushmere (Kent)
Guston (Kent)
GUYZANCE (Northumberland)
Gwaenysgor (Flintshire)
Gwalchmai (Anglesey)
GWAUN-CAE-GURWEN (Carmarthenshire)
Gwehelog (Monmouthshire)
Gwenddwr (Powys)
Gwernaffield (Flintshire)
Gwernesney (Monmouthshire)
Gwernogle (Carmarthenshire)
Gwernymynydd (Flintshire)
GWERSYLLT (Wrexham)
GWESPYR (Flintshire)
Gwredog (Anglesey)
GWRHAY (Caerphilly)
Gwyddelwern (Denbighshire)
Gwyddgrug (Carmarthenshire)
Gwynfryn (Wrexham)
Gwystre (Powys)
GYFELIA (Wrexham)

H

Habberley (Hereford & Worcester)
Habberley (Shropshire)
HABERGHAM (Lancashire)
Habertoft (Lincolnshire)
Habrough (Lincolnshire)
Hacconby (Lincolnshire)
Haceby (Lincolnshire)
Hack Green (Cheshire)
HACKENTHORPE (South Yorkshire)
Hackforth (North Yorkshire)
HACKLINGE (Kent)
Hackman's Gate (Hereford & Worcester)
Hackness (North Yorkshire)
Hackness (Somerset)
Hackthorn (Lincolnshire)
HACKTHORPE (Cumbria)
Haddington (Lincolnshire)
HADE EDGE (West Yorkshire)
HADFIELD (Derbyshire)
Hadley (Hereford & Worcester)
HADLEY (Shropshire)
Hadley End (Staffordshire)
Hadlow (Kent)
Hadnall (Shropshire)
Hadzor (Hereford & Worcester)
Haffenden Quarter (Kent)
HAFOD-Y-BWCH (Wrexham)
HAFOD-Y-COED (Blaenau Gwent)
HAFODYRYNYS (Caerphilly)
HAGGATE (Lancashire)
HAGGBECK (Cumbria)
HAGGERSTON (Northumberland)
Hagley (Hereford & Worcester)
Hagnaby (Lincolnshire)
Hagworthingham (Lincolnshire)
HAIGH (Greater Manchester)
Haighton Green (Lancashire)
Haile (Cumbria)
Hailes (Gloucestershire)
Haine (Kent)
Hainton (Lincolnshire)
HAINWORTH (West Yorkshire)
Haisthorpe (East Riding of Yorkshire)
Hakin (Pembrokeshire)
HALAM (Nottinghamshire)
Hale (Cheshire)
Hale (Cumbria)
Hale (Greater Manchester)
Hale (Somerset)
Hale Bank (Cheshire)
Hale Nook (Lancashire)
Hale Street (Kent)
Halebarns (Greater Manchester)
Hales (Staffordshire)
Hales Green (Derbyshire)
Hales Place (Kent)
Halesgate (Lincolnshire)
HALESOWEN (West Midlands)

Halewood (Merseyside)
Halewood Green (Merseyside)
Halford (Shropshire)
Halford (Warwickshire)
Halfpenny (Cumbria)
Halfpenny Green (Staffordshire)
Halfpenny Houses (North Yorkshire)
Halfway (Carmarthenshire)
HALFWAY (South Yorkshire)
HALFWAY HOUSE (Shropshire)
Halfway Houses (Kent)
HALIFAX (West Yorkshire)
Halkyn (Flintshire)
HALL CLIFFE (West Yorkshire)
Hall Cross (Lancashire)
Hall Dunnerdale (Cumbria)
HALL END (West Midlands)
Hall Green (West Midlands)
HALLAM FIELDS (Derbyshire)
Hallaton (Leicestershire)
HALLATROW (Somerset)
HALLBANKGATE (Cumbria)
Hallbeck (Cumbria)
HALLEN (Gloucestershire)
HALLFIELD GATE (Derbyshire)
HALLGARTH (Durham)
Halling (Kent)
Hallington (Lincolnshire)
HALLINGTON (Northumberland)
HALLIWELL (Greater Manchester)
Halloughton (Nottinghamshire)
Hallow (Hereford & Worcester)
Hallow Heath (Hereford & Worcester)
Hallthwaites (Cumbria)
Halltoft End (Lincolnshire)
HALMER END (Staffordshire)
Halmond's Frome (Hereford & Worcester)
Halsall (Lancashire)
Halse (Somerset)
Halsham (East Riding of Yorkshire)
Halstead (Kent)
Halstead (Leicestershire)
Halsway (Somerset)
Haltcliff Bridge (Cumbria)
Haltham (Lincolnshire)
Halton (Cheshire)
HALTON (Lancashire)
HALTON (Northumberland)
HALTON (West Yorkshire)
HALTON (Wrexham)
HALTON EAST (North Yorkshire)
Halton Fenside (Lincolnshire)
Halton Gill (North Yorkshire)
HALTON GREEN (Lancashire)
Halton Holegate (Lincolnshire)
HALTON LEA GATE (Northumberland)
HALTON SHIELDS (Northumberland)
Halton West (North Yorkshire)
HALTWHISTLE (Northumberland)
Ham (Gloucestershire)
HAM (Kent)
HAM (Somerset)
Ham Green (Hereford & Worcester)
Ham Green (Kent)
Ham Green (Somerset)
Ham Hill (Kent)
Ham Street (Somerset)
Hambleton (Lancashire)
HAMBLETON (North Yorkshire)
Hambleton Moss Side (Lancashire)
Hambridge (Somerset)
HAMBROOK (Gloucestershire)
Hameringham (Lincolnshire)
HAMMERWICH (Staffordshire)
Hampnett (Gloucestershire)
HAMPOLE (South Yorkshire)
Hampsfield (Cumbria)
Hampson Green (Lancashire)
HAMPSTHWAITE (North Yorkshire)
Hampton (Hereford & Worcester)
Hampton (Kent)
HAMPTON (Shropshire)
Hampton Bishop (Hereford & Worcester)
Hampton Green (Cheshire)
Hampton Heath (Cheshire)
Hampton in Arden (West Midlands)
HAMPTON LOADE (Shropshire)
Hampton Lovett (Hereford & Worcester)
Hampton Lucy (Warwickshire)
Hampton on the Hill (Warwickshire)
HAMSTALL RIDWARE (Staffordshire)

HAMSTEAD (West Midlands)
HAMSTERLEY (Durham)
Hamstreet (Kent)
Hamwood (Somerset)
Hanbury (Hereford & Worcester)
Hanbury (Staffordshire)
Hanby (Lincolnshire)
HANCHURCH (Staffordshire)
Hand Green (Cheshire)
Handale (North Yorkshire)
Handbridge (Cheshire)
Handforth (Cheshire)
Handley (Cheshire)
HANDLEY (Derbyshire)
HANDSACRE (Staffordshire)
HANDSWORTH (South Yorkshire)
HANDSWORTH (West Midlands)
HANFORD (Staffordshire)
HANHAM (Gloucestershire)
Hankelow (Cheshire)
HANLEY (Staffordshire)
Hanley Castle (Hereford & Worcester)
Hanley Child (Hereford & Worcester)
Hanley Swan (Hereford & Worcester)
Hanley William (Hereford & Worcester)
Hanlith (North Yorkshire)
HANMER (Wrexham)
Hannah (Lincolnshire)
Hanthorpe (Lincolnshire)
Hapsford (Cheshire)
HAPTON (Lancashire)
Harbledown (Kent)
Harborne (West Midlands)
Harborough Magna (Warwickshire)
Harborough Parva (Warwickshire)
Harbottle (Northumberland)
Harbours Hill (Hereford & Worcester)
Harbury (Warwickshire)
HARBY (Leicestershire)
Harby (Nottinghamshire)
HARDEN (West Midlands)
HARDEN (West Yorkshire)
Hardgate (North Yorkshire)
Hardhorn (Lancashire)
HARDINGS WOOD (Staffordshire)
HARDINGTON (Somerset)
Hardington Mandeville (Somerset)
Hardington Marsh (Somerset)
Hardington Moor (Somerset)
HARDRAW (North Yorkshire)
HARDSOUGH (Lancashire)
HARDSTOFT (Derbyshire)
Hardway (Somerset)
Hardwick (Lincolnshire)
HARDWICK (South Yorkshire)
HARDWICK (West Midlands)
Hardwick Green (Hereford & Worcester)
Hardwicke (Gloucestershire)
HARE CROFT (West Yorkshire)
Hareby (Lincolnshire)
Harehill (Derbyshire)
HAREHILLS (West Yorkshire)
HAREHOPE (Northumberland)
HARELAW (Durham)
Hareplain (Kent)
HARESCEUGH (Cumbria)
Harescombe (Gloucestershire)
Haresfield (Gloucestershire)
Harewood (West Yorkshire)
Harewood End (Hereford & Worcester)
Hargrave (Cheshire)
Harker (Cumbria)
HARLASTON (Staffordshire)
Harlaxton (Lincolnshire)
HARLE SYKE (Lancashire)
HARLESCOTT (Shropshire)
HARLESTHORPE (Derbyshire)
Harley (Shropshire)
HARLEY (South Yorkshire)
HARLINGTON (South Yorkshire)
HARLOW HILL (Northumberland)
Harlthorpe (East Riding of Yorkshire)
HARMBY (North Yorkshire)
Harmer Hill (Shropshire)
Harmston (Lincolnshire)
Harnage (Shropshire)
Harnham (Northumberland)
Harnhill (Gloucestershire)
HAROLDSTON WEST (Pembrokeshire)
Harome (North Yorkshire)
Harpham (East Riding of Yorkshire)

See paras. 2 and 4 of the User Guide 2003 if you can't find your place name.

Harpley (Hereford & Worcester)
Harpswell (Lincolnshire)
Harpur Hill (Derbyshire)
HARPURHEY (Greater Manchester)
Harraby (Cumbria)
Harrietsham (Kent)
Harrington (Lincolnshire)
HARRISEAHEAD (Staffordshire)
HARRISTON (Cumbria)
Harrogate (North Yorkshire)
HARROP DALE (Greater Manchester)
Harrowgate Village (Durham)
HARSTON (Leicestershire)
Harswell (East Riding of Yorkshire)
HART (Durham)
HART STATION (Durham)
Hartburn (Northumberland)
Hartford (Cheshire)
Hartford (Somerset)
HARTFORTH (North Yorkshire)
Harthill (Cheshire)
HARTHILL (South Yorkshire)
Hartington (Derbyshire)
Hartington (Northumberland)
Hartlebury (Hereford & Worcester)
Hartlepool (Durham)
HARTLEY (Cumbria)
Hartley (Kent)
HARTLEY (Northumberland)
Hartley Green (Kent)
Hartley Green (Staffordshire)
Hartlip (Kent)
HARTOFT END (North Yorkshire)
Harton (North Yorkshire)
Harton (Shropshire)
HARTON (Tyne & Wear)
Hartpury (Gloucestershire)
HARTSHEAD (West Yorkshire)
HARTSHEAD MOOR SIDE (West Yorkshire)
HARTSHILL (Staffordshire)
HARTSHILL (Warwickshire)
HARTSHORNE (Derbyshire)
Hartside (Northumberland)
Hartsop (Cumbria)
Hartswell (Somerset)
HARTWITH (North Yorkshire)
Harvel (Kent)
Harvington (Hereford & Worcester)
HARWELL (Nottinghamshire)
Harwood (Durham)
HARWOOD (Greater Manchester)
Harwood Dale (North Yorkshire)
HARWOOD LEE (Greater Manchester)
HARWORTH (Nottinghamshire)
HASBURY (West Midlands)
Haselbury Plucknett (Somerset)
Haseley (Warwickshire)
Haseley Green (Warwickshire)
Haseley Knob (Warwickshire)
Haselor (Warwickshire)
Hasfield (Gloucestershire)
Hasguard (Pembrokeshire)
Haskayne (Lancashire)
HASLAND (Derbyshire)
HASLAND GREEN (Derbyshire)
HASLINGDEN (Lancashire)
HASLINGDEN GRANE (Lancashire)
Haslington (Cheshire)
Hassall (Cheshire)
Hassall Green (Cheshire)
Hassell Street (Kent)
Hassness (Cumbria)
Hassop (Derbyshire)
Hasthorpe (Lincolnshire)
Hastingleigh (Kent)
Hastings (Somerset)
HASWELL (Durham)
HASWELL PLOUGH (Durham)
Hatch Beauchamp (Somerset)
Hatchmere (Cheshire)
Hatcliffe (Lincolnshire)
Hatfield (Hereford & Worcester)
HATFIELD (South Yorkshire)
HATFIELD WOODHOUSE (South Yorkshire)
Hathern (Leicestershire)
Hatherop (Gloucestershire)
Hathersage (Derbyshire)
Hathersage Booths (Derbyshire)
Hatherton (Cheshire)
HATHERTON (Staffordshire)
Hatton (Cheshire)

Hatton (Derbyshire)
Hatton (Lincolnshire)
Hatton (Shropshire)
Hatton (Warwickshire)
Hatton Heath (Cheshire)
Haugh (Lincolnshire)
Haugh (West Yorkshire)
Haugh Head (Northumberland)
Haugham (Lincolnshire)
HAUGHTON (Nottinghamshire)
Haughton (Powys)
HAUGHTON (Shropshire)
Haughton (Staffordshire)
HAUGHTON GREEN (Greater Manchester)
Haughton le Skerne (Durham)
Haughton Moss (Cheshire)
HAUNTON (Staffordshire)
HAUXLEY (Northumberland)
Havannah (Cheshire)
Haven (Hereford & Worcester)
Haven Bank (Lincolnshire)
Haven Side (East Riding of Yorkshire)
HAVERCROFT (West Yorkshire)
Haverfordwest (Pembrokeshire)
Haverigg (Cumbria)
Haverthwaite (Cumbria)
Haverton Hill (Durham)
HAVYAT GREEN (Somerset)
Havyatt (Somerset)
HAWARDEN (Flintshire)
Hawbridge (Hereford & Worcester)
Hawcoat (Cumbria)
Hawes (North Yorkshire)
Hawford (Hereford & Worcester)
HAWK GREEN (Greater Manchester)
Hawkenbury (Kent)
HAWKES END (West Midlands)
Hawkesbury (Gloucestershire)
HAWKESBURY (Warwickshire)
Hawkesbury Upton (Gloucestershire)
HAWKHILL (Northumberland)
Hawkhurst (Kent)
Hawkinge (Kent)
Hawksdale (Cumbria)
HAWKSHAW (Greater Manchester)
Hawkshead (Cumbria)
Hawkshead Hill (Cumbria)
Hawkstone (Shropshire)
HAWKSWICK (North Yorkshire)
Hawksworth (Nottinghamshire)
HAWKSWORTH (West Yorkshire)
Hawley (Kent)
Hawling (Gloucestershire)
HAWNBY (North Yorkshire)
HAWORTH (West Yorkshire)
HAWTHORN (Durham)
HAWTHORN (Rhondda Cynon Taff)
Hawthorn Hill (Lincolnshire)
Hawthorpe (Lincolnshire)
Hawton (Nottinghamshire)
Haxby (North Yorkshire)
Haxby Gates (North Yorkshire)
Haxey (Lincolnshire)
Haxey Turbary (Lincolnshire)
Hay-on-Wye (Powys)
HAYDOCK (Merseyside)
Haydon (Somerset)
HAYDON BRIDGE (Northumberland)
HAYFIELD (Derbyshire)
HAYGATE (Shropshire)
HAYLEY GREEN (West Midlands)
Haymoor Green (Cheshire)
Hayscastle (Pembrokeshire)
Hayscastle Cross (Pembrokeshire)
Haysden (Kent)
HAYTON (Cumbria)
Hayton (East Riding of Yorkshire)
HAYTON (Nottinghamshire)
Hayton's Bent (Shropshire)
Haywood (Hereford & Worcester)
HAYWOOD (South Yorkshire)
HAYWOOD OAKS (Nottinghamshire)
HAZEL GROVE (Greater Manchester)
Hazel Street (Kent)
Hazelford (Nottinghamshire)
HAZELHURST (Greater Manchester)
HAZELSLADE (Staffordshire)
Hazelwood (Derbyshire)
HAZLERIGG (Tyne & Wear)
HAZLES (Staffordshire)
Hazleton (Gloucestershire)

Headbrook (Hereford & Worcester)
Headcorn (Kent)
HEADINGLEY (West Yorkshire)
Headlam (Durham)
Headless Cross (Hereford & Worcester)
Headley Heath (Hereford & Worcester)
HEADON (Nottinghamshire)
Heads Nook (Cumbria)
HEAGE (Derbyshire)
Healaugh (North Yorkshire)
Heald Green (Greater Manchester)
Heale (Somerset)
HEALEY (Lancashire)
Healey (North Yorkshire)
HEALEY (Northumberland)
HEALEY (West Yorkshire)
HEALEYFIELD (Durham)
Healing (Lincolnshire)
HEANOR (Derbyshire)
HEAPEY (Lancashire)
Heapham (Lincolnshire)
Hearts Delight (Kent)
HEATH (Derbyshire)
HEATH (West Yorkshire)
HEATH END (Leicestershire)
Heath End (Warwickshire)
Heath Green (Hereford & Worcester)
HEATH HAYES (Staffordshire)
HEATH HILL (Shropshire)
Heath House (Somerset)
HEATH TOWN (West Midlands)
Heathbrook (Shropshire)
Heathcote (Derbyshire)
Heathcote (Shropshire)
HEATHER (Leicestershire)
HEATHFIELD (North Yorkshire)
Heathfield (Somerset)
Heathton (Shropshire)
Heatley (Greater Manchester)
Heatley (Staffordshire)
HEATON (Greater Manchester)
Heaton (Lancashire)
Heaton (Staffordshire)
HEATON (Tyne & Wear)
HEATON (West Yorkshire)
Heaton Chapel (Greater Manchester)
Heaton Mersey (Greater Manchester)
Heaton Norris (Greater Manchester)
Heaton's Bridge (Lancashire)
Heaverham (Kent)
HEAVILEY (Greater Manchester)
HEBBURN (Tyne & Wear)
HEBDEN (North Yorkshire)
HEBDEN BRIDGE (West Yorkshire)
Hebden Green (Cheshire)
Hebron (Anglesey)
Hebron (Carmarthenshire)
HEBRON (Northumberland)
Heckington (Lincolnshire)
HECKMONDWIKE (West Yorkshire)
HEDDON-ON-THE-WALL (Northumberland)
Hedging (Somerset)
HEDLEY ON THE HILL (Northumberland)
HEDNESFORD (Staffordshire)
Hedon (East Riding of Yorkshire)
Hegdon Hill (Hereford & Worcester)
Heighington (Durham)
Heighington (Lincolnshire)
HEIGHTINGTON (Hereford & Worcester)
Hele (Somerset)
HELLABY (South Yorkshire)
Hellifield (North Yorkshire)
HELM (Northumberland)
HELME (West Yorkshire)
HELMINGTON ROW (Durham)
HELMSHORE (Lancashire)
Helmsley (North Yorkshire)
Helmswell Cliff (Lincolnshire)
Helperby (North Yorkshire)
Helperthorpe (North Yorkshire)
Helpringham (Lincolnshire)
Helsby (Cheshire)
Helsey (Lincolnshire)
Helton (Cumbria)
HELWITH (North Yorkshire)
Helwith Bridge (North Yorkshire)
HEMINGBROUGH (North Yorkshire)
Hemingby (Lincolnshire)
HEMINGFIELD (South Yorkshire)
HEMINGTON (Somerset)
Hemlington (North Yorkshire)

See paras. 2 and 4 of the User Guide 2003 if you can't find your place name.

Hempholme (East Riding of Yorkshire)
Hempstead (Gloucestershire)
Hempstead (Kent)
Hemswell (Lincolnshire)
HEMSWORTH (West Yorkshire)
Henbury (Bristol)
Henbury (Cheshire)
Hendomen (Powys)
HENDRE (Bridgend)
HENDY (Carmarthenshire)
Heneglwys (Anglesey)
Henghurst (Kent)
HENGOED (Caerphilly)
Hengoed (Powys)
HENGOED (Shropshire)
Hengrove (Bristol)
Henhurst (Kent)
Heniarth (Powys)
Henlade (Somerset)
Henleaze (Bristol)
Henley (Gloucestershire)
Henley (Shropshire)
Henley (Somerset)
HENLEY GREEN (West Midlands)
Henley Street (Kent)
Henley-in-Arden (Warwickshire)
Henllan (Denbighshire)
Henllan Amgoed (Carmarthenshire)
Henllys (Torfaen)
Henry's Moat (Castell Hendre) (Pembrokeshire)
HENSALL (North Yorkshire)
HENSHAW (Northumberland)
HENSINGHAM (Cumbria)
Henstridge (Somerset)
Henstridge Ash (Somerset)
Henstridge Marsh (Somerset)
Henton (Somerset)
Henwick (Hereford & Worcester)
Heol Senni (Powys)
HEOL-LAS (Swansea)
HEOL-Y-CYW (Bridgend)
Hepburn (Northumberland)
HEPPLE (Northumberland)
HEPSCOTT (Northumberland)
HEPTONSTALL (West Yorkshire)
HEPWORTH (West Yorkshire)
Herbrandston (Pembrokeshire)
Hereford (Hereford & Worcester)
Hereson (Kent)
HERMIT HILL (South Yorkshire)
Hermon (Anglesey)
Hermon (Pembrokeshire)
Herne (Kent)
Herne Bay (Kent)
Herne Common (Kent)
Herne Pound (Kent)
Hernhill (Kent)
HERONDEN (Kent)
HERRINGTHORPE (South Yorkshire)
HERRINGTON (Tyne & Wear)
HERSDEN (Kent)
Hesketh Bank (Lancashire)
Hesketh Lane (Lancashire)
HESKIN GREEN (Lancashire)
HESLEDEN (Durham)
Hesleden (North Yorkshire)
HESLEYSIDE (Northumberland)
Heslington (North Yorkshire)
Hessay (North Yorkshire)
Hessle (East Riding of Yorkshire)
HESSLE (West Yorkshire)
Hest Bank (Lancashire)
Heswall (Merseyside)
Hethersgill (Cumbria)
Hetherside (Cumbria)
Hetherson Green (Cheshire)
Hethpool (Northumberland)
HETT (Durham)
HETTON (North Yorkshire)
HETTON STEADS (Northumberland)
HETTON-LE-HOLE (Tyne & Wear)
HEUGH (Northumberland)
Hever (Kent)
Heversham (Cumbria)
Hewelsfield (Gloucestershire)
HEWENDEN (West Yorkshire)
Hewish (Somerset)
HEXHAM (Northumberland)
Hextable (Kent)
HEXTHORPE (South Yorkshire)
Hey (Lancashire)

Hey Houses (Lancashire)
Heydour (Lincolnshire)
Heyhead (Greater Manchester)
HEYROD (Greater Manchester)
Heysham (Lancashire)
HEYSHAW (North Yorkshire)
HEYSIDE (Greater Manchester)
HEYWOOD (Greater Manchester)
Hibaldstow (Lincolnshire)
HICKLETON (South Yorkshire)
HICKLING (Nottinghamshire)
Hickmans Green (Kent)
HICKS FORSTAL (Kent)
Hidcote Bartrim (Gloucestershire)
Hidcote Boyce (Gloucestershire)
HIGH ACKWORTH (West Yorkshire)
HIGH ANGERTON (Northumberland)
High Bankhill (Cumbria)
HIGH BENTHAM (North Yorkshire)
High Bewaldeth (Cumbria)
High Bickwith (North Yorkshire)
HIGH BIGGINS (Cumbria)
High Borrans (Cumbria)
HIGH BRADLEY (North Yorkshire)
High Brooms (Kent)
HIGH BUSTON (Northumberland)
HIGH CALLERTON (Northumberland)
HIGH CASTERTON (Cumbria)
High Catton (East Riding of Yorkshire)
High Close (North Yorkshire)
High Coniscliffe (Durham)
High Crosby (Cumbria)
High Cross (Warwickshire)
HIGH CROSS BANK (Derbyshire)
HIGH DISLEY (Cheshire)
HIGH DUBMIRE (Tyne & Wear)
HIGH EGGBOROUGH (North Yorkshire)
High Ellington (North Yorkshire)
High Ercall (Shropshire)
HIGH ETHERLEY (Durham)
High Ferry (Lincolnshire)
HIGH FLATS (West Yorkshire)
HIGH GRANGE (Durham)
HIGH GRANTLEY (North Yorkshire)
High Green (Cumbria)
High Green (Hereford & Worcester)
HIGH GREEN (Shropshire)
HIGH GREEN (South Yorkshire)
HIGH GREEN (West Yorkshire)
High Halden (Kent)
High Halstow (Kent)
High Ham (Somerset)
HIGH HARRINGTON (Cumbria)
High Harrogate (North Yorkshire)
HIGH HASWELL (Durham)
HIGH HATTON (Shropshire)
High Hawsker (North Yorkshire)
High Hesket (Cumbria)
HIGH HOYLAND (South Yorkshire)
High Hunsley (East Riding of Yorkshire)
High Hutton (North Yorkshire)
High Ireby (Cumbria)
HIGH KILBURN (North Yorkshire)
High Killerby (North Yorkshire)
High Knipe (Cumbria)
HIGH LANDS (Durham)
High Lane (Cheshire)
HIGH LANE (Greater Manchester)
High Lane (Hereford & Worcester)
High Legh (Cheshire)
High Leven (North Yorkshire)
HIGH LITTLETON (Somerset)
High Lorton (Cumbria)
High Marnham (Nottinghamshire)
HIGH MELTON (South Yorkshire)
HIGH MICKLEY (Northumberland)
HIGH MOORSLEY (Tyne & Wear)
HIGH NEWPORT (Tyne & Wear)
High Newton (Cumbria)
HIGH NEWTON BY THE SEA (Northumberland)
High Nibthwaite (Cumbria)
High Offley (Staffordshire)
HIGH ONN (Staffordshire)
High Row (Cumbria)
HIGH SALTER (Lancashire)
HIGH SCALES (Cumbria)
HIGH SEATON (Cumbria)
HIGH SHAW (North Yorkshire)
High Side (Cumbria)
HIGH SPEN (Tyne & Wear)
HIGH STOOP (Durham)

High Street (Kent)
High Throston (Durham)
HIGH TOWN (Staffordshire)
High Toynton (Lincolnshire)
High Trewhitt (Northumberland)
HIGH URPETH (Durham)
HIGH WARDEN (Northumberland)
HIGH WESTWOOD (Durham)
High Woolaston (Gloucestershire)
High Worsall (North Yorkshire)
High Wray (Cumbria)
HIGHAM (Derbyshire)
Higham (Kent)
HIGHAM (Lancashire)
HIGHAM (South Yorkshire)
Higham Dykes (Northumberland)
Higham on the Hill (Leicestershire)
Highbridge (Somerset)
HIGHBURTON (West Yorkshire)
HIGHBURY (Somerset)
Highclifflane (Derbyshire)
Higher Alham (Somerset)
Higher Ballam (Lancashire)
Higher Bartle (Lancashire)
Higher Burwardsley (Cheshire)
Higher Chillington (Somerset)
Higher Combe (Somerset)
HIGHER HARPERS (Lancashire)
Higher Heysham (Lancashire)
HIGHER HURDSFIELD (Cheshire)
Higher Irlam (Greater Manchester)
HIGHER KINNERTON (Flintshire)
HIGHER OGDEN (Greater Manchester)
Higher Penwortham (Lancashire)
Higher Studfold (North Yorkshire)
Higher Walton (Cheshire)
Higher Walton (Lancashire)
Higher Wambrook (Somerset)
HIGHER WHEELTON (Lancashire)
Higher Whitley (Cheshire)
Higher Wych (Cheshire)
HIGHERFORD (Lancashire)
HIGHFIELD (East Riding of Yorkshire)
HIGHFIELD (Tyne & Wear)
HIGHFIELDS (South Yorkshire)
HIGHGATE (North Yorkshire)
HIGHGATE HEAD (Derbyshire)
HIGHGREEN MANOR (Northumberland)
HIGHLANE (South Yorkshire)
Highlaws (Cumbria)
Highleadon (Gloucestershire)
HIGHLEY (Shropshire)
HIGHMOOR (Cumbria)
Highmoor Hill (Monmouthshire)
Highnam (Gloucestershire)
Highnam Green (Gloucestershire)
HIGHRIDGE (Somerset)
Highstead (Kent)
Highsted (Kent)
Highter's Heath (West Midlands)
Hightown (Cheshire)
Hightown (Merseyside)
Highway (Hereford & Worcester)
Highwood (Staffordshire)
Hilden Park (Kent)
Hildenborough (Kent)
Hilderstone (Staffordshire)
Hilderthorpe (East Riding of Yorkshire)
Hill (Gloucestershire)
Hill (Warwickshire)
HILL CHORLTON (Staffordshire)
Hill Common (Somerset)
Hill Dyke (Lincolnshire)
HILL END (Durham)
Hill End (Gloucestershire)
Hill Green (Kent)
HILL RIDWARE (Staffordshire)
Hill Side (Hereford & Worcester)
HILL SIDE (West Yorkshire)
Hill Top (Durham)
HILL TOP (South Yorkshire)
HILL TOP (West Midlands)
HILL TOP (West Yorkshire)
HILLAM (North Yorkshire)
HILLBECK (Durham)
Hillborough (Kent)
Hillend (Swansea)
HILLERSLAND (Gloucestershire)
Hillesley (Gloucestershire)
Hillfarrance (Somerset)
Hillhampton (Hereford & Worcester)

See paras. 2 and 4 of the User Guide 2003 if you can't find your place name.

Hilliard's Cross (Staffordshire)
Hillmorton (Warwickshire)
HILLOCK VALE (Lancashire)
Hillpool (Hereford & Worcester)
HILLS TOWN (Derbyshire)
Hillside (Tyne & Wear)
Hilston (East Riding of Yorkshire)
Hilston Park (Monmouthshire)
HILTON (Cumbria)
Hilton (Derbyshire)
HILTON (Durham)
Hilton (North Yorkshire)
Hilton (Shropshire)
Himbleton (Hereford & Worcester)
HIMLEY (Staffordshire)
Hincaster (Cumbria)
Hinckley (Leicestershire)
Hinderwell (North Yorkshire)
Hindford (Shropshire)
HINDLE FOLD (Lancashire)
HINDLEY (Greater Manchester)
HINDLEY (Northumberland)
HINDLEY GREEN (Greater Manchester)
Hindlip (Hereford & Worcester)
Hinksford (Staffordshire)
HINNINGTON (Shropshire)
Hinstock (Shropshire)
Hinton (Gloucestershire)
Hinton (Hereford & Worcester)
HINTON (Shropshire)
HINTON BLEWETT (Somerset)
Hinton Charterhouse (Somerset)
Hinton Green (Hereford & Worcester)
Hinton on the Green (Hereford & Worcester)
Hinton St. George (Somerset)
HINTS (Shropshire)
Hints (Staffordshire)
Hinxhill (Kent)
HIPPERHOLME (West Yorkshire)
HIPSBURN (Northumberland)
Hipswell (North Yorkshire)
Hirnant (Powys)
HIRST (Northumberland)
HIRST COURTNEY (North Yorkshire)
Hirwaen (Denbighshire)
HIRWAUN (Rhondda Cynon Taff)
HISCOTT (Devon)
Hive (East Riding of Yorkshire)
Hixon (Staffordshire)
Hoaden (Kent)
Hoar Cross (Staffordshire)
Hoarwithy (Hereford & Worcester)
HOATH (Kent)
Hoathly (Kent)
Hobarris (Shropshire)
HOBSICK (Nottinghamshire)
HOBSON (Durham)
HOBY (Leicestershire)
Hoccombe (Somerset)
Hockerton (Nottinghamshire)
HOCKLEY (Cheshire)
HOCKLEY (Staffordshire)
HOCKLEY (West Midlands)
Hockley Heath (West Midlands)
HODDLESTON (Lancashire)
Hodgehill (Cheshire)
Hodgeston (Pembrokeshire)
Hodnet (Shropshire)
Hodsall Street (Kent)
HODSOCK (Nottinghamshire)
HODTHORPE (Derbyshire)
Hoff (Cumbria)
Hogben's Hill (Kent)
HOGHTON (Lancashire)
HOGHTON BOTTOMS (Lancashire)
Hognaston (Derbyshire)
HOGRILL'S END (Warwickshire)
Hogsthorpe (Lincolnshire)
Holbeach (Lincolnshire)
Holbeach Bank (Lincolnshire)
Holbeach Clough (Lincolnshire)
Holbeach Drove (Lincolnshire)
Holbeach Hurn (Lincolnshire)
Holbeach St. Johns (Lincolnshire)
Holbeach St. Mark's (Lincolnshire)
Holbeach St. Matthew (Lincolnshire)
HOLBECK (Nottinghamshire)
HOLBECK WOODHOUSE (Nottinghamshire)
Holberrow Green (Hereford & Worcester)
Holborough (Kent)
HOLBROOK (Derbyshire)

HOLBROOK (South Yorkshire)
HOLBROOK MOOR (Derbyshire)
HOLBURN (Northumberland)
HOLCOMBE (Greater Manchester)
HOLCOMBE (Somerset)
HOLCOMBE BROOK (Greater Manchester)
Holden (Lancashire)
HOLDEN GATE (West Yorkshire)
Holdgate (Shropshire)
Holdingham (Lincolnshire)
HOLDSWORTH (West Yorkshire)
Hole-in-the-Wall (Hereford & Worcester)
HOLEHOUSE (Derbyshire)
Holford (Somerset)
Holgate (North Yorkshire)
Holker (Cumbria)
Hollam (Somerset)
Holland Fen (Lincolnshire)
HOLLAND LEES (Lancashire)
Hollies Hill (Hereford & Worcester)
Hollin Green (Cheshire)
Hollingbourne (Kent)
HOLLINGTHORPE (West Yorkshire)
Hollington (Derbyshire)
Hollington (Staffordshire)
HOLLINGWORTH (Greater Manchester)
Hollinlane (Cheshire)
HOLLINS (Derbyshire)
HOLLINS (Greater Manchester)
HOLLINS (Staffordshire)
HOLLINS END (South Yorkshire)
Hollins Green (Cheshire)
Hollins Lane (Lancashire)
Hollinsclough (Staffordshire)
HOLLINSWOOD (Shropshire)
Hollinwood (Shropshire)
Holloway (Derbyshire)
Hollowmoor Heath (Cheshire)
Holly Green (Hereford & Worcester)
HOLLYBUSH (Caerphilly)
Hollybush (Hereford & Worcester)
Hollyhurst (Cheshire)
Hollym (East Riding of Yorkshire)
Hollywood (Hereford & Worcester)
Holmbridge (West Yorkshire)
Holmcroft (Staffordshire)
Holme (Cumbria)
Holme (Lincolnshire)
Holme (North Yorkshire)
HOLME (Nottinghamshire)
Holme (West Yorkshire)
HOLME CHAPEL (Lancashire)
HOLME GREEN (North Yorkshire)
Holme Lacy (Hereford & Worcester)
Holme Marsh (Hereford & Worcester)
Holme on the Wolds (East Riding of Yorkshire)
Holme Pierrepont (Nottinghamshire)
Holme St. Cuthbert (Cumbria)
Holme upon Spalding Moor (East Riding of Yorkshire)
Holmer (Hereford & Worcester)
Holmes Chapel (Cheshire)
HOLMESFIELD (Derbyshire)
Holmeswood (Lancashire)
HOLMEWOOD (Derbyshire)
HOLMFIELD (West Yorkshire)
HOLMFIRTH (West Yorkshire)
HOLMGATE (Derbyshire)
Holmpton (East Riding of Yorkshire)
Holmrook (Cumbria)
HOLMSIDE (Durham)
Holmwrangle (Cumbria)
Holnicote (Somerset)
Holt (Hereford & Worcester)
HOLT (Wrexham)
Holt End (Hereford & Worcester)
Holt Fleet (Hereford & Worcester)
Holt Green (Lancashire)
Holt Heath (Hereford & Worcester)
Holt Street (Kent)
Holtby (North Yorkshire)
Holton (Somerset)
Holton cum Beckering (Lincolnshire)
Holton le Clay (Lincolnshire)
Holton le Moor (Lincolnshire)
Holway (Flintshire)
HOLWELL (Leicestershire)
Holwick (Durham)
Holy Cross (Hereford & Worcester)
Holy Island (Northumberland)
Holyhead (Anglesey)

HOLYMOORSIDE (Derbyshire)
HOLYSTONE (Northumberland)
HOLYWELL (Flintshire)
HOLYWELL GREEN (West Yorkshire)
Holywell Lake (Somerset)
Hom Green (Hereford & Worcester)
Homer (Shropshire)
Homer Green (Merseyside)
Homescales (Cumbria)
Honey Hill (Kent)
Honeyborough (Pembrokeshire)
Honeybourne (Hereford & Worcester)
Honiley (Warwickshire)
Honington (Lincolnshire)
Honington (Warwickshire)
HONLEY (West Yorkshire)
HONNINGTON (Shropshire)
Hoo (Kent)
Hoo Green (Cheshire)
Hoobrook (Hereford & Worcester)
HOOD GREEN (South Yorkshire)
HOOD HILL (South Yorkshire)
Hoohill (Lancashire)
Hook (East Riding of Yorkshire)
Hook (Kent)
HOOK (Pembrokeshire)
Hook Bank (Hereford & Worcester)
Hook Green (Kent)
Hook Street (Gloucestershire)
HOOKAGATE (Shropshire)
Hookgate (Staffordshire)
HOOLEY BRIDGE (Greater Manchester)
Hooton (Cheshire)
HOOTON LEVITT (South Yorkshire)
HOOTON PAGNELL (South Yorkshire)
HOOTON ROBERTS (South Yorkshire)
Hop Pole (Lincolnshire)
Hope (Derbyshire)
HOPE (Flintshire)
Hope (Powys)
HOPE (Shropshire)
Hope (Staffordshire)
Hope Bowdler (Shropshire)
Hope Mansell (Hereford & Worcester)
Hope under Dinmore (Hereford & Worcester)
Hopesay (Shropshire)
HOPETOWN (West Yorkshire)
Hopperton (North Yorkshire)
Hopsford (Warwickshire)
HOPSTONE (Shropshire)
Hopton (Derbyshire)
Hopton (Shropshire)
Hopton (Staffordshire)
Hopton Cangeford (Shropshire)
Hopton Castle (Shropshire)
HOPTON WAFERS (Shropshire)
Hoptonheath (Shropshire)
Hopwas (Staffordshire)
HOPWOOD (Greater Manchester)
Hopwood (Hereford & Worcester)
Horbling (Lincolnshire)
HORBURY (West Yorkshire)
Horcott (Gloucestershire)
HORDEN (Durham)
Horderley (Shropshire)
Hordley (Shropshire)
HOREB (Carmarthenshire)
HORFIELD (Bristol)
Horkstow (Lincolnshire)
Horn Street (Kent)
Hornblotton Green (Somerset)
HORNBY (Lancashire)
Hornby (North Yorkshire)
Horncastle (Lincolnshire)
Horner (Somerset)
Horninghold (Leicestershire)
Horninglow (Staffordshire)
Hornsby (Cumbria)
Hornsby (Somerset)
Hornsbygate (Cumbria)
Hornsea (East Riding of Yorkshire)
HORROCKS FOLD (Greater Manchester)
Horrocksford (Lancashire)
HORSEBRIDGE (Shropshire)
HORSEBRIDGE (Staffordshire)
Horsebrook (Staffordshire)
Horsecastle (Somerset)
Horsegate (Lincolnshire)
HORSEHAY (Shropshire)
HORSEHOUSE (North Yorkshire)
HORSEMAN'S GREEN (Wrexham)

Horsey (Somerset)
HORSFORTH (West Yorkshire)
Horsham (Hereford & Worcester)
Horsington (Lincolnshire)
Horsington (Somerset)
HORSLEY (Derbyshire)
Horsley (Gloucestershire)
Horsley (Northumberland)
HORSLEY WOODHOUSE (Derbyshire)
HORSLEY-GATE (Derbyshire)
Horsmonden (Kent)
Horton (Gloucestershire)
Horton (Lancashire)
HORTON (Shropshire)
Horton (Somerset)
Horton (Staffordshire)
Horton (Swansea)
Horton Cross (Somerset)
Horton Green (Cheshire)
Horton in Ribblesdale (North Yorkshire)
Horton Kirby (Kent)
HORWICH (Greater Manchester)
HORWICH END (Derbyshire)
HOSCAR (Lancashire)
HOSE (Leicestershire)
Hosey Hill (Kent)
Hotham (East Riding of Yorkshire)
Hothfield (Kent)
Hoton (Leicestershire)
HOTT (Northumberland)
HOTWELLS (Bristol)
Hough (Cheshire)
HOUGH END (West Yorkshire)
HOUGH GREEN (Cheshire)
Hough-on-the-Hill (Lincolnshire)
Hougham (Lincolnshire)
Houghton (Cumbria)
HOUGHTON (Northumberland)
Houghton (Pembrokeshire)
HOUGHTON GREEN (Cheshire)
Houghton le Side (Durham)
HOUGHTON LE SPRING (Tyne & Wear)
Houghton on the Hill (Leicestershire)
Houndsmoor (Somerset)
HOUSES HILL (West Yorkshire)
HOVE EDGE (West Yorkshire)
HOVERINGHAM (Nottinghamshire)
Hovingham (North Yorkshire)
How (Cumbria)
How Caple (Hereford & Worcester)
HOWBROOK (South Yorkshire)
Howden (East Riding of Yorkshire)
HOWDEN-LE-WEAR (Durham)
Howe (North Yorkshire)
HOWE BRIDGE (Greater Manchester)
Howell (Lincolnshire)
Howey (Powys)
Howgill (North Yorkshire)
HOWICK (Northumberland)
Howle (Durham)
Howle (Shropshire)
Howle Hill (Hereford & Worcester)
Howley (Somerset)
HOWRIGG (Cumbria)
Howsham (Lincolnshire)
Howsham (North Yorkshire)
Howt Green (Kent)
Howtel (Northumberland)
Howton (Hereford & Worcester)
Howtown (Cumbria)
Hoylake (Merseyside)
HOYLAND COMMON (South Yorkshire)
HOYLAND NETHER (South Yorkshire)
HOYLAND SWAINE (South Yorkshire)
HOYLE MILL (South Yorkshire)
HUBBERHOLME (North Yorkshire)
Hubberston (Pembrokeshire)
Hubbert's Bridge (Lincolnshire)
Huby (North Yorkshire)
Hucclecote (Gloucestershire)
Hucking (Kent)
HUCKNALL (Nottinghamshire)
HUDDERSFIELD (West Yorkshire)
Huddington (Hereford & Worcester)
HUDSWELL (North Yorkshire)
Huggate (East Riding of Yorkshire)
HUGGLESCOTE (Leicestershire)
Hughley (Shropshire)
Huish Champflower (Somerset)
Huish Episcopi (Somerset)
Hulberry (Kent)

Hull (East Riding of Yorkshire)
Hulland (Derbyshire)
Hulland Ward (Derbyshire)
HULME (Cheshire)
Hulme (Greater Manchester)
HULME (Staffordshire)
Hulme End (Staffordshire)
Hulme Walfield (Cheshire)
Hulse Heath (Cheshire)
HULTON LANE ENDS (Greater Manchester)
Humberston (Lincolnshire)
Humberstone (Leicestershire)
Humberton (North Yorkshire)
Humbleton (East Riding of Yorkshire)
Humbleton (Northumberland)
Humby (Lincolnshire)
Humshaugh (Northumberland)
HUNCOAT (Lancashire)
Huncote (Leicestershire)
HUNDALL (Derbyshire)
Hunderthwaite (Durham)
Hundle Houses (Lincolnshire)
Hundleby (Lincolnshire)
Hundleton (Pembrokeshire)
Hundred End (Lancashire)
Hundred House (Powys)
Hundred The (Hereford & Worcester)
Hungarton (Leicestershire)
HUNGER HILL (Lancashire)
Hungerford (Somerset)
Hungerstone (Hereford & Worcester)
Hungerton (Lincolnshire)
Hungryhatton (Shropshire)
Hunmanby (North Yorkshire)
Hunningham (Warwickshire)
Hunnington (Hereford & Worcester)
Hunsingore (North Yorkshire)
HUNSLET (West Yorkshire)
Hunsonby (Cumbria)
Hunstanworth (Durham)
HUNSTRETE (Somerset)
HUNSWORTH (West Yorkshire)
Hunt End (Hereford & Worcester)
Hunt's Cross (Merseyside)
Hunterston (Cheshire)
Huntham (Somerset)
Huntingdon (Hereford & Worcester)
Huntington (Hereford & Worcester)
Huntington (North Yorkshire)
HUNTINGTON (Staffordshire)
Huntley (Gloucestershire)
Hunton (Kent)
Hunton (North Yorkshire)
Hunts Green (Warwickshire)
Huntscott (Somerset)
Huntspill (Somerset)
Huntstile (Somerset)
Huntworth (Somerset)
HUNWICK (Durham)
Hurcott (Somerset)
HURDSFIELD (Cheshire)
HURLEY (Warwickshire)
HURLEY COMMON (Warwickshire)
Hurlston Green (Lancashire)
Hurn's End (Lincolnshire)
HURST (North Yorkshire)
Hurst (Somerset)
Hurst Green (Lancashire)
HURST HILL (West Midlands)
Hurstley (Hereford & Worcester)
Hurstway Common (Hereford & Worcester)
HURSTWOOD (Lancashire)
HURWORTH BURN (Durham)
Hurworth-on-Tees (Durham)
Hury (Durham)
Husbands Bosworth (Leicestershire)
Husthwaite (North Yorkshire)
HUT GREEN (North Yorkshire)
HUTHWAITE (North Yorkshire)
HUTHWAITE (Nottinghamshire)
Huttoft (Lincolnshire)
Hutton (Cumbria)
Hutton (East Riding of Yorkshire)
Hutton (Lancashire)
Hutton (Somerset)
Hutton Bonville (North Yorkshire)
Hutton Buscel (North Yorkshire)
Hutton Conyers (North Yorkshire)
Hutton Cranswick (East Riding of Yorkshire)
Hutton End (Cumbria)
Hutton Hall (North Yorkshire)

Hutton Hang (North Yorkshire)
HUTTON HENRY (Durham)
HUTTON LOWCROSS (North Yorkshire)
Hutton Magna (Durham)
HUTTON MULGRAVE (North Yorkshire)
Hutton Roof (Cumbria)
Hutton Rudby (North Yorkshire)
HUTTON SESSAY (North Yorkshire)
Hutton Wandesley (North Yorkshire)
Hutton-le-Hole (North Yorkshire)
Huxham Green (Somerset)
Huxley (Cheshire)
HUYTON (Merseyside)
Hycemoor (Cumbria)
Hyde (Gloucestershire)
HYDE (Greater Manchester)
Hyde Lea (Staffordshire)
Hyde Park Corner (Somerset)
Hykeham Moor (Lincolnshire)
Hyssington (Powys)
Hystfield (Gloucestershire)
Hythe (Kent)
Hythe (Somerset)
Hyton (Cumbria)

I

Ible (Derbyshire)
IBSTOCK (Leicestershire)
Iburndale (North Yorkshire)
Icelton (Somerset)
ICKHAM (Kent)
Ickornshaw (North Yorkshire)
Icomb (Gloucestershire)
Ide Hill (Kent)
Iden Green (Kent)
IDLE (West Yorkshire)
Idlicote (Warwickshire)
Idridgehay (Derbyshire)
Ifton (Monmouthshire)
IFTON HEATH (Shropshire)
Ightam (Kent)
Ightfield (Shropshire)
Ilam (Staffordshire)
Ilchester (Somerset)
Ilderton (Northumberland)
Ilford (Somerset)
ILKESTON (Derbyshire)
ILKLEY (West Yorkshire)
Illey (West Midlands)
Illidge Green (Cheshire)
ILLINGWORTH (West Yorkshire)
Illston on the Hill (Leicestershire)
Ilmington (Warwickshire)
Ilminster (Somerset)
ILSTON (Swansea)
ILTON (North Yorkshire)
Ilton (Somerset)
Immingham (Lincolnshire)
Immingham Dock (Lincolnshire)
Ince (Cheshire)
Ince Blundell (Merseyside)
INCE-IN-MAKERFIELD (Greater Manchester)
INGBIRCHWORTH (South Yorkshire)
Ingerthorpe (North Yorkshire)
Ingestre (Staffordshire)
Ingham (Lincolnshire)
INGLEBY (Derbyshire)
Ingleby Arncliffe (North Yorkshire)
Ingleby Barwick (North Yorkshire)
Ingleby Cross (North Yorkshire)
Ingleby Greenhow (North Yorkshire)
Inglesbatch (Somerset)
Ingleton (Durham)
INGLETON (North Yorkshire)
Inglewhite (Lancashire)
Ingmire Hall (Cumbria)
INGOE (Northumberland)
Ingoldmells (Lincolnshire)
Ingoldsby (Lincolnshire)
Ingram (Northumberland)
INGROW (West Yorkshire)
Ings (Cumbria)
Ingst (Gloucestershire)
Ingthorpe (Lincolnshire)
Inkberrow (Hereford & Worcester)
INKERMAN (Durham)
Inskip (Lancashire)
Inskip Moss Side (Lancashire)
INTAKE (South Yorkshire)
IPSTONES (Staffordshire)

See paras. 2 and 4 of the User Guide 2003 if you can't find your place name.

Irby (Merseyside)
Irby in the Marsh (Lincolnshire)
Irby upon Humber (Lincolnshire)
IREBY (Cumbria)
IREBY (Lancashire)
Ireleth (Cumbria)
Ireshopeburn (Durham)
Ireton Wood (Derbyshire)
Irlam (Greater Manchester)
Irnham (Lincolnshire)
IRON ACTON (Gloucestershire)
Iron Cross (Warwickshire)
IRONBRIDGE (Shropshire)
IRONVILLE (Derbyshire)
Irthington (Cumbria)
Irton (North Yorkshire)
Islandpool (Hereford & Worcester)
Isle Abbotts (Somerset)
Isle Brewers (Somerset)
Isley Walton (Leicestershire)
Isombridge (Shropshire)
Istead Rise (Kent)
Itchington (Gloucestershire)
Itton (Monmouthshire)
IVEGILL (Cumbria)
Ivelet (North Yorkshire)
IVESTON (Durham)
Ivington (Hereford & Worcester)
Ivington Green (Hereford & Worcester)
Ivy Hatch (Kent)
Ivychurch (Kent)
Iwade (Kent)

J

Jack Green (Lancashire)
Jack Hill (North Yorkshire)
JACKSDALE (Nottinghamshire)
JACKSON BRIDGE (West Yorkshire)
Jameston (Pembrokeshire)
JARROW (Tyne & Wear)
Jeator Houses (North Yorkshire)
JEFFRESTON (Pembrokeshire)
Jerusalem (Lincolnshire)
JESMOND (Tyne & Wear)
Jingle Street (Monmouthshire)
Jodrell Bank (Cheshire)
Johnby (Cumbria)
JOHNSTON (Carmarthenshire)
JOHNSTON (Pembrokeshire)
JOHNSTOWN (Wrexham)
Jordanston (Pembrokeshire)
JORDANTHORPE (South Yorkshire)
Joyden's Wood (Kent)
Jubilee Corner (Kent)
JUMP (South Yorkshire)

K

KABER (Cumbria)
Keadby (Lincolnshire)
Keal Cotes (Lincolnshire)
Kearby Town End (North Yorkshire)
KEARSLEY (Greater Manchester)
KEARSLEY (Northumberland)
Kearsney (Kent)
Kearstwick (Cumbria)
Kearton (North Yorkshire)
KEASDEN (North Yorkshire)
Keckwick (Cheshire)
Keddington (Lincolnshire)
Keddington Corner (Lincolnshire)
Kedleston (Derbyshire)
Keelby (Lincolnshire)
KEELE (Staffordshire)
KEELE UNIVERSITY (Staffordshire)
KEELHAM (West Yorkshire)
KEESTON (Pembrokeshire)
Kegworth (Leicestershire)
KEIGHLEY (West Yorkshire)
Keinton Mandeville (Somerset)
KEIRSLEYWELL ROW (Northumberland)
Keisby (Lincolnshire)
KEISLEY (Cumbria)
Kelbrook (Lancashire)
Kelby (Lincolnshire)
Keld (Cumbria)
KELD (North Yorkshire)
Keld Head (North Yorkshire)
Keldholme (North Yorkshire)
Kelfield (Lincolnshire)

KELFIELD (North Yorkshire)
Kelham (Nottinghamshire)
Kellamergh (Lancashire)
Kelleth (Cumbria)
KELLINGTON (North Yorkshire)
KELLOE (Durham)
KELLS (Cumbria)
Kelsall (Cheshire)
Kelsick (Cumbria)
KELSTEDGE (Derbyshire)
Kelstern (Lincolnshire)
KELSTERTON (Flintshire)
Kelston (Somerset)
KEMBERTON (Shropshire)
Kemble (Gloucestershire)
Kemble Wick (Gloucestershire)
Kemerton (Hereford & Worcester)
Kemeys Commander (Monmouthshire)
Kempe's Corner (Kent)
Kempley (Gloucestershire)
Kempley Green (Gloucestershire)
Kemps Green (Warwickshire)
Kempsey (Hereford & Worcester)
Kempsford (Gloucestershire)
Kempton (Shropshire)
Kemsing (Kent)
Kemsley (Kent)
Kemsley Street (Kent)
Kenardington (Kent)
Kenchester (Hereford & Worcester)
Kendal (Cumbria)
Kenderchurch (Hereford & Worcester)
KENDLESHIRE (Gloucestershire)
Kenfig (Bridgend)
KENFIG HILL (Bridgend)
Kenilworth (Warwickshire)
Kenley (Shropshire)
Kenn (Somerset)
Kennessee Green (Merseyside)
Kennington (Kent)
Kenny (Somerset)
Kennythorpe (North Yorkshire)
Kensham Green (Kent)
Kensington (Merseyside)
KENT GREEN (Cheshire)
Kent Street (Kent)
Kent's Green (Gloucestershire)
Kentchurch (Hereford & Worcester)
Kentmere (Cumbria)
KENTON (Tyne & Wear)
KENTON BANK FOOT (Northumberland)
Kents Bank (Cumbria)
Kenwick (Shropshire)
KENYON (Cheshire)
Kepwick (North Yorkshire)
KERESLEY (West Midlands)
KERESLEY GREEN (Warwickshire)
Kerne Bridge (Hereford & Worcester)
KERRIDGE (Cheshire)
KERRIDGE-END (Cheshire)
Kerry (Powys)
KERSALL (Nottinghamshire)
Kersoe (Hereford & Worcester)
Kerswell Green (Hereford & Worcester)
Keswick (Cumbria)
Ketsby (Lincolnshire)
KETTLEBROOK (Staffordshire)
KETTLESHULME (Cheshire)
KETTLESING (North Yorkshire)
KETTLESING BOTTOM (North Yorkshire)
Kettlethorpe (Lincolnshire)
KETTLEWELL (North Yorkshire)
KEXBROUGH (South Yorkshire)
Kexby (Lincolnshire)
Kexby (North Yorkshire)
KEY GREEN (Cheshire)
KEY GREEN (North Yorkshire)
Key Street (Kent)
Key's Toft (Lincolnshire)
Keyham (Leicestershire)
Keyingham (East Riding of Yorkshire)
KEYNSHAM (Somerset)
KEYWORTH (Nottinghamshire)
Kibbear (Somerset)
KIBBLESWORTH (Tyne & Wear)
Kibworth Beauchamp (Leicestershire)
Kibworth Harcourt (Leicestershire)
KIDBURNGILL (Cumbria)
Kiddemore Green (Staffordshire)
Kidderminster (Hereford & Worcester)
KIDSGROVE (Staffordshire)

KIDSTONES (North Yorkshire)
KIDWELLY (Carmarthenshire)
Kielder (Northumberland)
KILBURN (Derbyshire)
KILBURN (North Yorkshire)
Kilby (Leicestershire)
KILCOT (Gloucestershire)
KILDALE (North Yorkshire)
KILDWICK (North Yorkshire)
Kilford (Denbighshire)
KILGETTY (Pembrokeshire)
Kilgwrrwg Common (Monmouthshire)
Kilham (East Riding of Yorkshire)
Kilham (Northumberland)
KILLAMARSH (Derbyshire)
KILLAY (Swansea)
Killerby (Durham)
Killinghall (North Yorkshire)
Killington (Cumbria)
KILLINGWORTH (Tyne & Wear)
KILMERSDON (Somerset)
KILN PIT HILL (Northumberland)
Kilndown (Kent)
Kilnhill (Cumbria)
Kilnhouses (Cheshire)
KILNHURST (South Yorkshire)
Kilnsea (East Riding of Yorkshire)
KILNSEY (North Yorkshire)
Kilnwick (East Riding of Yorkshire)
Kilnwick Percy (East Riding of Yorkshire)
Kilpeck (Hereford & Worcester)
Kilpin (East Riding of Yorkshire)
Kilpin Pike (East Riding of Yorkshire)
Kilrie (Cheshire)
Kilton (North Yorkshire)
Kilton Thorpe (North Yorkshire)
Kilve (Somerset)
Kilvington (Nottinghamshire)
KIMBERLEY (Nottinghamshire)
KIMBERWORTH (South Yorkshire)
KIMBLESWORTH (Durham)
Kimbolton (Hereford & Worcester)
Kimcote (Leicestershire)
Kimmerston (Northumberland)
Kineton (Gloucestershire)
Kineton (Warwickshire)
King Sterndale (Derbyshire)
King's Acre (Hereford & Worcester)
KING'S BROMLEY (Staffordshire)
King's Coughton (Warwickshire)
King's Heath (West Midlands)
King's Hill (Warwickshire)
KING'S MOSS (Lancashire)
KING'S NEWTON (Derbyshire)
King's Norton (Leicestershire)
King's Norton (West Midlands)
King's Pyon (Hereford & Worcester)
King's Stanley (Gloucestershire)
Kingcoed (Monmouthshire)
Kingerby (Lincolnshire)
KINGS BRIDGE (Swansea)
Kings Caple (Hereford & Worcester)
Kings Green (Gloucestershire)
KINGS HILL (West Midlands)
KINGS MEABURN (Cumbria)
Kings Newnham (Warwickshire)
Kings Weston (Bristol)
Kingsbridge (Somerset)
KINGSBURY (Warwickshire)
Kingsbury Episcopi (Somerset)
Kingscote (Gloucestershire)
Kingsdon (Somerset)
Kingsdown (Bristol)
Kingsdown (Kent)
KINGSFORD (Hereford & Worcester)
Kingsgate (Kent)
Kingshurst (West Midlands)
Kingside Hill (Cumbria)
Kingsland (Anglesey)
Kingsland (Hereford & Worcester)
Kingsley (Cheshire)
KINGSLEY (Staffordshire)
Kingslow (Shropshire)
Kingsnorth (Kent)
KINGSTANDING (West Midlands)
Kingsthorne (Hereford & Worcester)
Kingston (Kent)
KINGSTON ON SOAR (Nottinghamshire)
Kingston Seymour (Somerset)
Kingston St. Mary (Somerset)
Kingstone (Hereford & Worcester)

See paras. 2 and 4 of the User Guide 2003 if you can't find your place name.

Kingstone (Somerset)
Kingstone (Staffordshire)
Kingstown (Cumbria)
KINGSWINFORD (West Midlands)
KINGSWOOD (Bristol / Gloucestershire)
Kingswood (Kent)
Kingswood (Powys)
Kingswood (Somerset)
Kingswood (Warwickshire)
Kingswood Brook (Warwickshire)
Kingswood Common (Hereford & Worcester)
Kingswood Common (Staffordshire)
Kingthorpe (Lincolnshire)
Kington (Gloucestershire)
Kington (Hereford & Worcester)
Kingweston (Somerset)
KINLET (Shropshire)
Kinnerley (Shropshire)
Kinnersley (Hereford & Worcester)
Kinnerton (Powys)
Kinnerton (Shropshire)
KINNERTON GREEN (Flintshire)
Kinninvie (Durham)
KINOULTON (Nottinghamshire)
Kinsey Heath (Cheshire)
Kinsham (Hereford & Worcester)
KINSLEY (West Yorkshire)
Kinton (Hereford & Worcester)
Kinton (Shropshire)
Kinver (Staffordshire)
Kiplin (North Yorkshire)
KIPPAX (West Yorkshire)
Kipping's Cross (Kent)
Kirby Bellars (Leicestershire)
KIRBY CORNER (West Midlands)
Kirby Fields (Leicestershire)
Kirby Grindalythe (North Yorkshire)
KIRBY HILL (North Yorkshire)
Kirby Knowle (North Yorkshire)
Kirby Misperton (North Yorkshire)
Kirby Muxloe (Leicestershire)
Kirby Sigston (North Yorkshire)
Kirby Underdale (East Riding of Yorkshire)
Kirby Wiske (North Yorkshire)
KIRK BRAMWITH (South Yorkshire)
Kirk Deighton (North Yorkshire)
Kirk Ella (East Riding of Yorkshire)
KIRK HALLAM (Derbyshire)
Kirk Hammerton (North Yorkshire)
Kirk Ireton (Derbyshire)
Kirk Langley (Derbyshire)
KIRK MERRINGTON (Durham)
KIRK SANDALL (South Yorkshire)
KIRK SMEATON (North Yorkshire)
Kirkandrews upon Eden (Cumbria)
Kirkbampton (Cumbria)
Kirkbride (Cumbria)
Kirkbridge (North Yorkshire)
Kirkburn (East Riding of Yorkshire)
KIRKBURTON (West Yorkshire)
Kirkby (Lincolnshire)
Kirkby (Merseyside)
Kirkby (North Yorkshire)
Kirkby Fleetham (North Yorkshire)
Kirkby Green (Lincolnshire)
Kirkby Hall (North Yorkshire)
KIRKBY IN ASHFIELD (Nottinghamshire)
Kirkby la Thorpe (Lincolnshire)
KIRKBY LONSDALE (Cumbria)
Kirkby Malham (North Yorkshire)
Kirkby Mallory (Leicestershire)
KIRKBY MALZEARD (North Yorkshire)
Kirkby Mills (North Yorkshire)
Kirkby on Bain (Lincolnshire)
Kirkby Overblow (North Yorkshire)
KIRKBY STEPHEN (Cumbria)
Kirkby Thore (Cumbria)
Kirkby Underwood (Lincolnshire)
Kirkby Wharf (North Yorkshire)
KIRKBY WOODHOUSE (Nottinghamshire)
Kirkby-in-Furness (Cumbria)
Kirkbymoorside (North Yorkshire)
Kirkcambeck (Cumbria)
Kirkdale (Merseyside)
Kirkham (Lancashire)
Kirkham (North Yorkshire)
KIRKHAMGATE (West Yorkshire)
KIRKHARLE (Northumberland)
KIRKHAUGH (Northumberland)
KIRKHEATON (Northumberland)
KIRKHEATON (West Yorkshire)

KIRKHOUSE (Cumbria)
KIRKHOUSE GREEN (South Yorkshire)
KIRKLAND (Cumbria)
KIRKLAND GUARDS (Cumbria)
Kirkleatham (North Yorkshire)
Kirklevington (North Yorkshire)
Kirklington (North Yorkshire)
KIRKLINGTON (Nottinghamshire)
Kirklinton (Cumbria)
Kirknewton (Northumberland)
Kirkoswald (Cumbria)
Kirksanton (Cumbria)
KIRKSTALL (West Yorkshire)
Kirkstead (Lincolnshire)
Kirkstone Pass Inn (Cumbria)
KIRKTHORPE (West Yorkshire)
KIRKWHELPINGTON (Northumberland)
Kirmington (Lincolnshire)
Kirmond le Mire (Lincolnshire)
Kirton (Lincolnshire)
KIRTON (Nottinghamshire)
Kirton End (Lincolnshire)
Kirton Holme (Lincolnshire)
Kirton in Lindsey (Lincolnshire)
Kite Green (Warwickshire)
Kitebrook (Warwickshire)
Kites Hardwick (Warwickshire)
KITT GREEN (Greater Manchester)
Kittisford (Somerset)
KITTLE (Swansea)
Kitts Green (West Midlands)
Kivernoll (Hereford & Worcester)
KIVETON PARK (South Yorkshire)
Knaith (Lincolnshire)
Knaith Park (Lincolnshire)
Knaplock (Somerset)
Knapp (Somerset)
KNAPTHORPE (Nottinghamshire)
Knapton (North Yorkshire)
Knapton Green (Hereford & Worcester)
Knaresborough (North Yorkshire)
KNARSDALE (Northumberland)
Knayton (North Yorkshire)
Knedlington (East Riding of Yorkshire)
KNEESALL (Nottinghamshire)
Kneeton (Nottinghamshire)
Knelston (Swansea)
KNENHALL (Staffordshire)
Knightcote (Warwickshire)
Knightley (Staffordshire)
Knightley Dale (Staffordshire)
Knighton (Leicestershire)
Knighton (Powys)
Knighton (Somerset)
Knighton (Staffordshire)
Knighton on Teme (Hereford & Worcester)
Knightsbridge (Gloucestershire)
Knightwick (Hereford & Worcester)
Knill (Hereford & Worcester)
KNIPTON (Leicestershire)
KNITSLEY (Durham)
Kniveton (Derbyshire)
Knock (Cumbria)
Knockhall (Kent)
Knockholt (Kent)
Knockholt Pound (Kent)
Knockin (Shropshire)
Knockmill (Kent)
Knole (Somerset)
KNOLE PARK (Gloucestershire)
Knolls Green (Cheshire)
KNOLTON (Wrexham)
Knossington (Leicestershire)
Knott End-on-Sea (Lancashire)
KNOTTINGLEY (West Yorkshire)
Knotty Ash (Merseyside)
KNOWBURY (Shropshire)
KNOWLE (Bristol)
KNOWLE (Shropshire)
Knowle (Somerset)
Knowle (West Midlands)
Knowle Green (Lancashire)
Knowle St. Giles (Somerset)
Knowlefield (Cumbria)
KNOWLTON (Kent)
KNOWSLEY (Merseyside)
KNOWSLEY HALL (Merseyside)
Knox (North Yorkshire)
Knox Bridge (Kent)
Knucklas (Powys)
Knutsford (Cheshire)

KNUTTON (Staffordshire)
KNYPERSLEY (Staffordshire)
Krumlin (West Yorkshire)
KYLOE (Northumberland)
Kynaston (Hereford & Worcester)
Kynaston (Shropshire)
Kynnersley (Shropshire)
Kyre Green (Hereford & Worcester)
Kyre Park (Hereford & Worcester)
Kyrewood (Hereford & Worcester)
Kyrle (Somerset)

L

Laceby (Lincolnshire)
Lach Dennis (Cheshire)
Lackenby (North Yorkshire)
Ladbroke (Warwickshire)
Ladderedge (Staffordshire)
Laddingford (Kent)
Lade Bank (Lincolnshire)
Lady Hall (Cumbria)
Ladyridge (Hereford & Worcester)
Ladywood (Hereford & Worcester)
Ladywood (West Midlands)
LAISTERDYKE (West Yorkshire)
Laithes (Cumbria)
Lake Side (Cumbria)
Laleston (Bridgend)
Lamb Roe (Lancashire)
Lamberhurst (Kent)
Lamberhurst Down (Kent)
LAMBLEY (Northumberland)
LAMBLEY (Nottinghamshire)
Lambston (Pembrokeshire)
LAMESLEY (Tyne & Wear)
LAMONBY (Cumbria)
Lampeter Velfrey (Pembrokeshire)
Lamphey (Pembrokeshire)
LAMPLUGH (Cumbria)
Lamyatt (Somerset)
Lancaster (Lancashire)
Lancaut (Gloucestershire)
LANCHESTER (Durham)
Landimore (Swansea)
LANDORE (Swansea)
LANDSHIPPING (Pembrokeshire)
LANDYWOOD (Staffordshire)
LANE BOTTOM (Lancashire)
LANE END (Cheshire)
Lane End (Kent)
Lane End (Lancashire)
Lane End Waberthwaite (Cumbria)
Lane Ends (Derbyshire)
LANE ENDS (Durham)
LANE ENDS (Lancashire)
Lane Ends (North Yorkshire)
Lane Green (Staffordshire)
Lane Head (Durham)
LANE HEAD (Greater Manchester)
LANE HEAD (West Midlands)
Lane Heads (Lancashire)
LANE SIDE (Lancashire)
Laneham (Nottinghamshire)
Lanehead (Durham)
Lanehead (Northumberland)
LANESHAW BRIDGE (Lancashire)
Langaller (Somerset)
LANGAR (Nottinghamshire)
LANGBAR (North Yorkshire)
Langbaurgh (North Yorkshire)
Langcliffe (North Yorkshire)
Langdale End (North Yorkshire)
Langdon Beck (Durham)
LANGFORD (Nottinghamshire)
Langford (Somerset)
Langford Budville (Somerset)
LANGHO (Lancashire)
Langland (Swansea)
LANGLEY (Cheshire)
LANGLEY (Derbyshire)
Langley (Gloucestershire)
LANGLEY (Greater Manchester)
Langley (Kent)
LANGLEY (Northumberland)
Langley (Somerset)
Langley (Warwickshire)
LANGLEY CASTLE (Northumberland)
Langley Common (Derbyshire)
Langley Green (Derbyshire)
Langley Green (Warwickshire)

See paras. 2 and 4 of the User Guide 2003 if you can't find your place name.

Langley Marsh (Somerset)
LANGLEY MILL (Derbyshire)
LANGLEY MOOR (Durham)
LANGLEY PARK (Durham)
LANGOLD (Nottinghamshire)
Langport (Somerset)
Langrick (Lincolnshire)
Langridge (Somerset)
LANGRIGG (Cumbria)
LANGSETT (South Yorkshire)
Langstone (Newport)
Langthorne (North Yorkshire)
Langthorpe (North Yorkshire)
LANGTHWAITE (North Yorkshire)
Langtoft (East Riding of Yorkshire)
Langtoft (Lincolnshire)
Langton (Durham)
Langton (Lincolnshire)
Langton (North Yorkshire)
Langton by Wragby (Lincolnshire)
Langton Green (Kent)
Langwathby (Cumbria)
LANGWITH (Derbyshire)
Langworth (Lincolnshire)
Lanton (Northumberland)
Lapley (Staffordshire)
Lapworth (Warwickshire)
Larbreck (Lancashire)
Larkfield (Kent)
LARTINGTON (Durham)
Lasborough (Gloucestershire)
Lashenden (Kent)
Lask Edge (Staffordshire)
Lastingham (North Yorkshire)
Latcham (Somerset)
LATEBROOK (Staffordshire)
LATELY COMMON (Greater Manchester)
Latteridge (Gloucestershire)
Lattiford (Somerset)
Laugharne (Carmarthenshire)
Laughterton (Lincolnshire)
Laughton (Leicestershire)
Laughton (Lincolnshire)
LAUGHTON-EN-LE-MORTHEN (South Yorkshire)
Lavernock (Vale of Glamorgan)
Laversdale (Cumbria)
Laverton (Gloucestershire)
LAVERTON (North Yorkshire)
Laverton (Somerset)
LAVISTER (Wrexham)
Lawford (Somerset)
Lawkland (North Yorkshire)
Lawkland Green (North Yorkshire)
LAWLEY (Shropshire)
Lawnhead (Staffordshire)
Lawrence Weston (Bristol)
LAWRENNY (Pembrokeshire)
Lawton (Hereford & Worcester)
Laxton (East Riding of Yorkshire)
LAXTON (Nottinghamshire)
Laycock (West Yorkshire)
Laytham (East Riding of Yorkshire)
Laythes (Cumbria)
Lazenby (North Yorkshire)
Lazonby (Cumbria)
Lea (Derbyshire)
Lea (Hereford & Worcester)
Lea (Lincolnshire)
LEA (Shropshire)
Lea Bridge (Derbyshire)
Lea Heath (Staffordshire)
LEA MARSTON (Warwickshire)
Lea Town (Lancashire)
Lea Yeat (Cumbria)
Leadenham (Lincolnshire)
LEADGATE (Durham)
LEADGATE (Northumberland)
Leadingcross Green (Kent)
Leadmill (Derbyshire)
Leahead (Cheshire)
LEAHOLM SIDE (North Yorkshire)
Leake (North Yorkshire)
Leake Common Side (Lincolnshire)
Lealholm (North Yorkshire)
Leam (Derbyshire)
Leamington Hastings (Warwickshire)
Leamington Spa (Warwickshire)
Leamonsley (Staffordshire)
LEAMSIDE (Durham)
Leasgill (Cumbria)
Leasingham (Lincolnshire)

LEASINGTHORNE (Durham)
LEATHLEY (North Yorkshire)
Leaton (Shropshire)
Leaveland (Kent)
Leavening (North Yorkshire)
Lebberston (North Yorkshire)
Lechlade (Gloucestershire)
LECK (Lancashire)
Leckhampton (Gloucestershire)
Leckwith (Vale of Glamorgan)
Leconfield (East Riding of Yorkshire)
Ledbury (Hereford & Worcester)
Leddington (Gloucestershire)
Ledgemoor (Hereford & Worcester)
Ledicot (Hereford & Worcester)
Ledsham (Cheshire)
LEDSHAM (West Yorkshire)
LEDSTON (West Yorkshire)
LEDSTON LUCK (West Yorkshire)
Lee (Shropshire)
Lee Brockhurst (Shropshire)
Lee Green (Cheshire)
LEEBOTWOOD (Shropshire)
Leece (Cumbria)
Leeds (Kent)
LEEDS (West Yorkshire)
Leeds Beck (Lincolnshire)
Leek (Staffordshire)
Leek Wootton (Warwickshire)
Leeming (North Yorkshire)
LEEMING (West Yorkshire)
Leeming Bar (North Yorkshire)
Lees (Derbyshire)
LEES (Greater Manchester)
LEES (West Yorkshire)
Lees Green (Derbyshire)
Lees Hill (Cumbria)
Leesthorpe (Leicestershire)
LEESWOOD (Flintshire)
Leftwich (Cheshire)
Legbourne (Lincolnshire)
Legburthwaite (Cumbria)
Legsby (Lincolnshire)
Leicester (Leicestershire)
Leicester Forest East (Leicestershire)
Leigh (Gloucestershire)
LEIGH (Greater Manchester)
Leigh (Hereford & Worcester)
Leigh (Kent)
Leigh (Shropshire)
Leigh Green (Kent)
Leigh Sinton (Hereford & Worcester)
Leigh upon Mendip (Somerset)
LEIGH WOODS (Somerset)
Leighland Chapel (Somerset)
Leighterton (Gloucestershire)
LEIGHTON (North Yorkshire)
Leighton (Powys)
LEIGHTON (Shropshire)
Leighton (Somerset)
Leinthall Earls (Hereford & Worcester)
Leinthall Starkes (Hereford & Worcester)
Leintwardine (Hereford & Worcester)
Leire (Leicestershire)
Lelley (East Riding of Yorkshire)
LEM HILL (Hereford & Worcester)
LEMMINGTON HALL (Northumberland)
Lenchwick (Hereford & Worcester)
Lenham (Kent)
Lenham Heath (Kent)
Lenton (Lincolnshire)
LENTON (Nottinghamshire)
Leominster (Hereford & Worcester)
Leonard Stanley (Gloucestershire)
Leppington (North Yorkshire)
LEPTON (West Yorkshire)
LESBURY (Northumberland)
Lessonhall (Cumbria)
Lett's Green (Kent)
Letterston (Pembrokeshire)
Letton (Hereford & Worcester)
LETWELL (South Yorkshire)
Levedale (Staffordshire)
Leven (East Riding of Yorkshire)
Levens (Cumbria)
LEVENSHULME (Greater Manchester)
Leverton (Lincolnshire)
LEVISHAM (North Yorkshire)
Leweston (Pembrokeshire)
Lewis Wych (Hereford & Worcester)
Lewson Street (Kent)

Lewth (Lancashire)
Lexworthy (Somerset)
Leybourne (Kent)
LEYBURN (North Yorkshire)
LEYCETT (Staffordshire)
LEYLAND (Lancashire)
LEYLAND GREEN (Merseyside)
Leysdown-on-Sea (Kent)
Leysters (Hereford & Worcester)
Libanus (Powys)
Lichfield (Staffordshire)
Lickey (Hereford & Worcester)
Lickey End (Hereford & Worcester)
Lickey Rock (Hereford & Worcester)
LIDGATE (Derbyshire)
LIDGET (South Yorkshire)
LIDGETT (Nottinghamshire)
Lidsing (Kent)
Lifford (West Midlands)
Lighthazles (West Yorkshire)
Lighthorne (Warwickshire)
LIGHTWOOD (Staffordshire)
Lightwood Green (Cheshire)
LIGHTWOOD GREEN (Wrexham)
Lilburn Tower (Northumberland)
LILLESHALL (Shropshire)
Lilstock (Somerset)
LILYHURST (Shropshire)
LIMBRICK (Lancashire)
Lime Street (Hereford & Worcester)
Limebrook (Hereford & Worcester)
LIMEFIELD (Greater Manchester)
LIMESTONE BRAE (Northumberland)
Limington (Somerset)
LINBY (Nottinghamshire)
Lincoln (Lincolnshire)
Lincomb (Hereford & Worcester)
Lindal in Furness (Cumbria)
Lindale (Cumbria)
LINDLEY (West Yorkshire)
LINDLEY GREEN (North Yorkshire)
Lindow End (Cheshire)
Lindridge (Hereford & Worcester)
Liney (Somerset)
LINGBOB (West Yorkshire)
Lingdale (North Yorkshire)
Lingen (Hereford & Worcester)
LINGLEY GREEN (Cheshire)
Linkend (Hereford & Worcester)
Linkhill (Kent)
Linley (Shropshire)
Linley Green (Hereford & Worcester)
LINLEYGREEN (Shropshire)
Linshiels (Northumberland)
Linstock (Cumbria)
Linthurst (Hereford & Worcester)
Linthwaite (West Yorkshire)
LINTON (Derbyshire)
Linton (Hereford & Worcester)
Linton (Kent)
LINTON (North Yorkshire)
Linton (West Yorkshire)
LINTON HEATH (Derbyshire)
Linton Hill (Gloucestershire)
Linton-on-Ouse (North Yorkshire)
Linwood (Lincolnshire)
Lipley (Shropshire)
Liscard (Merseyside)
Liscombe (Somerset)
Lissett (East Riding of Yorkshire)
Lissington (Lincolnshire)
Lisvane (Cardiff)
Liswerry (Newport)
LITCHARD (Bridgend)
Litherland (Merseyside)
Little Alne (Warwickshire)
Little Altcar (Merseyside)
LITTLE ASBY (Cumbria)
Little Aston (Staffordshire)
Little Ayton (North Yorkshire)
Little Badminton (Gloucestershire)
Little Bampton (Cumbria)
Little Barrington (Gloucestershire)
Little Barrow (Cheshire)
Little Barugh (North Yorkshire)
Little Bavington (Northumberland)
LITTLE BAYTON (Warwickshire)
Little Birch (Hereford & Worcester)
Little Bispham (Lancashire)
Little Blencow (Cumbria)
LITTLE BLOXWICH (West Midlands)

Little Bolehill (Derbyshire)
Little Bowden (Leicestershire)
Little Brampton (Hereford & Worcester)
Little Brampton (Shropshire)
Little Bridgeford (Staffordshire)
LITTLE BROUGHTON (Cumbria)
Little Budworth (Cheshire)
Little Bytham (Lincolnshire)
Little Carlton (Lincolnshire)
Little Carlton (Nottinghamshire)
Little Catwick (East Riding of Yorkshire)
Little Cawthorpe (Lincolnshire)
Little Charlinch (Somerset)
Little Chart (Kent)
Little Cheveney (Kent)
LITTLE CLIFTON (Cumbria)
Little Coates (Lincolnshire)
Little Comberton (Hereford & Worcester)
Little Comp (Kent)
Little Compton (Warwickshire)
Little Corby (Cumbria)
Little Cowarne (Hereford & Worcester)
Little Crakehall (North Yorkshire)
Little Crosby (Merseyside)
Little Crosthwaite (Cumbria)
Little Cubley (Derbyshire)
Little Dalby (Leicestershire)
Little Dewchurch (Hereford & Worcester)
Little Doward (Hereford & Worcester)
Little Driffield (East Riding of Yorkshire)
Little Eaton (Derbyshire)
Little Elm (Somerset)
LITTLE FARINGDON (South Yorkshire)
Little Fencote (North Yorkshire)
LITTLE FENTON (North Yorkshire)
Little Garway (Hereford & Worcester)
Little Gorsley (Hereford & Worcester)
Little Green (Nottinghamshire)
LITTLE GREEN (Somerset)
Little Grimsby (Lincolnshire)
LITTLE GRINGLEY (Nottinghamshire)
Little Habton (North Yorkshire)
Little Hale (Lincolnshire)
LITTLE HALLAM (Derbyshire)
Little Hartlip (Kent)
Little Hatfield (East Riding of Yorkshire)
LITTLE HAVEN (Pembrokeshire)
Little Hay (Staffordshire)
LITTLE HAYFIELD (Derbyshire)
LITTLE HAYWOOD (Staffordshire)
Little Heath (Staffordshire)
LITTLE HEATH (West Midlands)
Little Hereford (Hereford & Worcester)
Little Hermitage (Kent)
LITTLE HORTON (West Yorkshire)
LITTLE HOUGHTON (South Yorkshire)
Little Hucklow (Derbyshire)
LITTLE HULTON (Greater Manchester)
LITTLE HUTTON (North Yorkshire)
Little Ingestre (Staffordshire)
Little Kelk (East Riding of Yorkshire)
Little Keyford (Somerset)
Little Kineton (Warwickshire)
Little Langdale (Cumbria)
Little Leigh (Cheshire)
LITTLE LEVER (Greater Manchester)
Little Load (Somerset)
Little London (Gloucestershire)
Little London (Lincolnshire)
Little London (Powys)
LITTLE LONDON (West Yorkshire)
Little Longstone (Derbyshire)
LITTLE MADELEY (Staffordshire)
Little Malvern (Hereford & Worcester)
LITTLE MANCOT (Flintshire)
Little Marcle (Hereford & Worcester)
Little Mill (Monmouthshire)
LITTLE MONGHAM (Kent)
Little Moor (Somerset)
Little Musgrave (Cumbria)
Little Ness (Shropshire)
LITTLE NESTON (Cheshire)
Little Newcastle (Pembrokeshire)
LITTLE NEWSHAM (Durham)
Little Norton (Somerset)
LITTLE NORTON (Staffordshire)
Little Onn (Staffordshire)
Little Ormside (Cumbria)
Little Orton (Cumbria)
Little Ouseburn (North Yorkshire)
LITTLE PACKINGTON (Warwickshire)

Little Pattenden (Kent)
Little Plumpton (Lancashire)
Little Ponton (Lincolnshire)
LITTLE PRESTON (West Yorkshire)
Little Reedness (East Riding of Yorkshire)
Little Ribston (North Yorkshire)
Little Rissington (Gloucestershire)
Little Rowsley (Derbyshire)
Little Ryle (Northumberland)
LITTLE RYTON (Shropshire)
Little Salkeld (Cumbria)
LITTLE SAREDON (Staffordshire)
Little Saughall (Cheshire)
LITTLE SESSAY (North Yorkshire)
Little Singleton (Lancashire)
LITTLE SKIPWITH (North Yorkshire)
LITTLE SMEATON (North Yorkshire)
Little Sodbury (Gloucestershire)
LITTLE SODBURY END (Gloucestershire)
Little Soudley (Shropshire)
Little Stainforth (North Yorkshire)
Little Stainton (Durham)
Little Stanney (Cheshire)
Little Steeping (Lincolnshire)
Little Stretton (Leicestershire)
Little Stretton (Shropshire)
LITTLE STRICKLAND (Cumbria)
Little Sugnall (Staffordshire)
Little Sutton (Cheshire)
Little Sutton (Shropshire)
LITTLE SWINBURNE (Northumberland)
LITTLE THIRKLEBY (North Yorkshire)
Little Thornton (Lancashire)
LITTLE THORPE (Durham)
LITTLE TOWN (Cheshire)
Little Town (Cumbria)
Little Town (Lancashire)
Little Twycross (Leicestershire)
Little Urswick (Cumbria)
Little Washbourne (Gloucestershire)
Little Weighton (East Riding of Yorkshire)
Little Welton (Lincolnshire)
LITTLE WENLOCK (Shropshire)
Little Weston (Somerset)
LITTLE WHITTINGTON (Northumberland)
Little Witcombe (Gloucestershire)
Little Witley (Hereford & Worcester)
Little Wolford (Warwickshire)
LITTLE WYRLEY (Staffordshire)
Little Wytheford (Shropshire)
LITTLEBECK (North Yorkshire)
LITTLEBOROUGH (Greater Manchester)
Littleborough (Nottinghamshire)
LITTLEBOURNE (Kent)
LITTLEDEAN (Gloucestershire)
LITTLEHARLE TOWER (Northumberland)
LITTLEHOUGHTON (Northumberland)
LITTLEMOOR (Derbyshire)
Littleover (Derbyshire)
Littler (Cheshire)
Littlestone-on-Sea (Kent)
Littlethorpe (Leicestershire)
Littlethorpe (North Yorkshire)
Littleton (Cheshire)
Littleton (Somerset)
Littleton-on-Severn (Gloucestershire)
LITTLETOWN (Durham)
LITTLEWOOD (Staffordshire)
Littleworth (Hereford & Worcester)
Littleworth (Staffordshire)
Litton (Derbyshire)
LITTON (North Yorkshire)
LITTON (Somerset)
Liverpool (City Centre) (Merseyside)
LIVERSEDGE (West Yorkshire)
Liverton (North Yorkshire)
Liverton Mines (North Yorkshire)
Liverton Street (Kent)
Lixwm (Flintshire)
Llaingoch (Anglesey)
Llaithddu (Powys)
Llan (Powys)
LLAN-Y-PWLL (Wrexham)
Llanafan-fechan (Powys)
Llanallgo (Anglesey)
Llanarmon-yn-Ial (Denbighshire)
Llanarth (Monmouthshire)
Llanarthne (Carmarthenshire)
LLANASA (Flintshire)
Llanbabo (Anglesey)
Llanbadarn Fynydd (Powys)

Llanbadarn-y-garreg (Powys)
Llanbadoc (Monmouthshire)
Llanbadrig (Anglesey)
Llanbeder (Newport)
Llanbedr (Powys)
Llanbedr-Dyffryn-Clwyd (Denbighshire)
Llanbedrgoch (Anglesey)
Llanbethery (Vale of Glamorgan)
Llanbister (Powys)
Llanblethian (Vale of Glamorgan)
Llanboidy (Carmarthenshire)
LLANBRADACH (Caerphilly)
Llanbrynmair (Powys)
Llancadle (Vale of Glamorgan)
Llancarfan (Vale of Glamorgan)
Llancayo (Monmouthshire)
Llancillo (Hereford & Worcester)
Llancloudy (Hereford & Worcester)
Llandaff (Cardiff)
Llandawke (Carmarthenshire)
Llanddanielfab (Anglesey)
Llanddarog (Carmarthenshire)
Llanddeusant (Anglesey)
Llanddeusant (Carmarthenshire)
Llanddew (Powys)
Llanddewi (Swansea)
Llanddewi Rhydderch (Monmouthshire)
Llanddewi Velfrey (Pembrokeshire)
Llanddewi Ystradenni (Powys)
Llanddewi'r Cwm (Powys)
Llanddona (Anglesey)
Llanddowror (Carmarthenshire)
Llanddyfnan (Anglesey)
Llandefaelog (Powys)
Llandefaelogtrer-graig (Powys)
Llandefalle (Powys)
Llandegfan (Anglesey)
Llandegla (Denbighshire)
Llandegley (Powys)
Llandegveth (Monmouthshire)
Llandeilo (Carmarthenshire)
Llandeilo Graban (Powys)
Llandeilo'r Fan (Powys)
Llandeloy (Pembrokeshire)
Llandenny (Monmouthshire)
Llandevaud (Newport)
Llandevenny (Monmouthshire)
Llandinabo (Hereford & Worcester)
Llandinam (Powys)
Llandissilio (Pembrokeshire)
Llandogo (Monmouthshire)
Llandough (Vale of Glamorgan)
Llandovery (Carmarthenshire)
Llandow (Vale of Glamorgan)
Llandre (Carmarthenshire)
Llandre Isaf (Pembrokeshire)
Llandrillo (Denbighshire)
Llandrindod Wells (Powys)
Llandrinio (Powys)
Llandulas (Powys)
LLANDYBIE (Carmarthenshire)
Llandyfaelog (Carmarthenshire)
LLANDYFAN (Carmarthenshire)
Llandyfrydog (Anglesey)
Llandynan (Denbighshire)
Llandyrnog (Denbighshire)
Llandyssil (Powys)
Llanedeyrn (Cardiff)
LLANEDI (Carmarthenshire)
Llaneglwys (Powys)
Llanegwad (Carmarthenshire)
Llaneilian (Anglesey)
Llanelidan (Denbighshire)
Llanelieu (Powys)
Llanellen (Monmouthshire)
LLANELLI (Carmarthenshire)
Llanelly (Monmouthshire)
Llanelwedd (Powys)
Llanerch (Powys)
Llanerchymedd (Anglesey)
Llanerfyl (Powys)
Llanfachraeth (Anglesey)
Llanfaelog (Anglesey)
Llanfaenor (Monmouthshire)
Llanfaes (Anglesey)
Llanfaes (Powys)
Llanfaethlu (Anglesey)
Llanfair Caereinion (Powys)
Llanfair Dyffryn Clwyd (Denbighshire)
Llanfair Kilgeddin (Monmouthshire)
Llanfair P G (Anglesey)

See paras. 2 and 4 of the User Guide 2003 if you can't find your place name.

51

Llanfair Waterdine (Shropshire)
Llanfair-Nant-Gwyn (Pembrokeshire)
Llanfair-y-Cwmmwd (Anglesey)
Llanfair-yn-Neubwll (Anglesey)
Llanfairynghornwy (Anglesey)
Llanfallteg (Carmarthenshire)
Llanfallteg West (Carmarthenshire)
Llanfechain (Powys)
Llanfechelli (Anglesey)
Llanferres (Denbighshire)
Llanfflewyn (Anglesey)
Llanfigael (Anglesey)
Llanfihangel Nant Bran (Powys)
Llanfihangel Rhydithon (Powys)
Llanfihangel Rogiet (Monmouthshire)
Llanfihangel Tal-y-llyn (Powys)
Llanfihangel yn Nhowyn (Anglesey)
Llanfihangel-ar-Arth (Carmarthenshire)
Llanfihangel-nant-Melan (Powys)
Llanfihangel-uwch-Gwili (Carmarthenshire)
Llanfihangel-yng-Ngwynfa (Powys)
Llanfilo (Powys)
Llanfoist (Monmouthshire)
Llanfrechfa (Torfaen)
Llanfrynach (Powys)
Llanfwrog (Anglesey)
Llanfwrog (Denbighshire)
Llanfyllin (Powys)
Llanfynydd (Carmarthenshire)
LLANFYNYDD (Flintshire)
Llanfyrnach (Pembrokeshire)
Llangadfan (Powys)
LLANGADOG (Carmarthenshire)
Llangadwaladr (Anglesey)
Llangadwaladr (Powys)
Llangaffo (Anglesey)
Llangain (Carmarthenshire)
Llangammarch Wells (Powys)
Llangan (Vale of Glamorgan)
Llangarron (Hereford & Worcester)
Llangasty-Talyllyn (Powys)
Llangathen (Carmarthenshire)
Llangattock (Powys)
Llangattock Lingoed (Monmouthshire)
Llangattock-Vibon-Avel (Monmouthshire)
Llangedwyn (Powys)
Llangefni (Anglesey)
LLANGEINOR (Bridgend)
Llangeinwen (Anglesey)
Llangeler (Carmarthenshire)
Llangendeirne (Carmarthenshire)
LLANGENNECH (Carmarthenshire)
Llangennith (Swansea)
Llangenny (Powys)
LLANGIWG (Neath Port Talbot)
Llangloffan (Pembrokeshire)
Llanglydwen (Carmarthenshire)
Llangoed (Anglesey)
Llangollen (Denbighshire)
Llangolman (Pembrokeshire)
Llangors (Powys)
Llangovan (Monmouthshire)
Llangristiolus (Anglesey)
Llangrove (Hereford & Worcester)
Llangua (Monmouthshire)
Llangunllo (Powys)
Llangunnor (Carmarthenshire)
Llangurig (Powys)
Llangwm (Monmouthshire)
LLANGWM (Pembrokeshire)
Llangwm-isaf (Monmouthshire)
Llangwyfan (Denbighshire)
Llangwyllog (Anglesey)
Llangybi (Monmouthshire)
LLANGYFELACH (Swansea)
Llangynhafal (Denbighshire)
Llangynidr (Powys)
Llangynin (Carmarthenshire)
Llangynog (Carmarthenshire)
Llangynog (Powys)
LLANGYNWYD (Bridgend)
Llanhamlach (Powys)
LLANHARAN (Rhondda Cynon Taff)
LLANHARRY (Rhondda Cynon Taff)
Llanhennock (Monmouthshire)
LLANHILLETH (Blaenau Gwent)
Llanidan (Anglesey)
Llanidloes (Powys)
Llanigon (Powys)
LLANILID (Rhondda Cynon Taff)
Llanishen (Cardiff)

Llanishen (Monmouthshire)
Llanlleonfel (Powys)
Llanllowell (Monmouthshire)
Llanllugan (Powys)
Llanllwch (Carmarthenshire)
Llanllwchaiarn (Powys)
Llanllwni (Carmarthenshire)
Llanmadoc (Swansea)
Llanmaes (Vale of Glamorgan)
Llanmartin (Newport)
Llanmerewig (Powys)
Llanmihangel (Vale of Glamorgan)
Llanmiloe (Carmarthenshire)
LLANMORLAIS (Swansea)
Llannon (Carmarthenshire)
Llanover (Monmouthshire)
Llanpumsaint (Carmarthenshire)
Llanrhaeadr-ym-Mochnant (Powys)
Llanrhidian (Swansea)
Llanrhyddlad (Anglesey)
Llanrian (Pembrokeshire)
Llanrothal (Hereford & Worcester)
Llanrumney (Cardiff)
Llansadurnen (Carmarthenshire)
Llansadwrn (Anglesey)
Llansadwrn (Carmarthenshire)
LLANSAINT (Carmarthenshire)
LLANSAMLET (Swansea)
Llansannor (Vale of Glamorgan)
Llansantffraed (Powys)
Llansantffraed-Cwmdeuddwr (Powys)
Llansantffraed-in-Elvel (Powys)
Llansantffraid-ym-Mechain (Powys)
Llansawel (Carmarthenshire)
Llansilin (Powys)
Llansoy (Monmouthshire)
Llanspyddid (Powys)
Llanstadwell (Pembrokeshire)
Llansteffan (Carmarthenshire)
Llanstephan (Powys)
Llantarnam (Torfaen)
Llanteg (Pembrokeshire)
Llanthewy Skirrid (Monmouthshire)
Llanthony (Monmouthshire)
Llantilio Pertholey (Monmouthshire)
Llantilio-Crossenny (Monmouthshire)
Llantrisant (Anglesey)
Llantrisant (Monmouthshire)
LLANTRISANT (Rhondda Cynon Taff)
Llantrithyd (Vale of Glamorgan)
LLANTWIT FARDRE (Rhondda Cynon Taff)
Llantwit Major (Vale of Glamorgan)
Llantysilio (Denbighshire)
Llanvaches (Newport)
Llanvair Discoed (Monmouthshire)
Llanvapley (Monmouthshire)
Llanvetherine (Monmouthshire)
Llanveynoe (Hereford & Worcester)
Llanvihangel Crucorney (Monmouthshire)
Llanvihangel Gobion (Monmouthshire)
Llanvihangel-Ystern-Llewern (Monmouthshire)
Llanwarne (Hereford & Worcester)
Llanwddyn (Powys)
Llanwenarth (Monmouthshire)
Llanwern (Newport)
Llanwinio (Carmarthenshire)
Llanwnda (Pembrokeshire)
Llanwnog (Powys)
LLANWONNO (Rhondda Cynon Taff)
Llanwrda (Carmarthenshire)
Llanwrin (Powys)
Llanwrthwl (Powys)
Llanwrtyd (Powys)
Llanwrtyd Wells (Powys)
Llanwyddelan (Powys)
Llanyblodwel (Shropshire)
Llanybri (Carmarthenshire)
Llanybydder (Carmarthenshire)
Llanycefn (Pembrokeshire)
Llanychaer Bridge (Pembrokeshire)
Llanycrwys (Carmarthenshire)
Llanymynech (Powys)
Llanynghenedl (Anglesey)
Llanynis (Powys)
Llanynys (Denbighshire)
Llanyre (Powys)
Llanywern (Powys)
Llawhaden (Pembrokeshire)
LLAWNT (Shropshire)
Llawryglyn (Powys)
LLAY (Wrexham)

Llechcynfarwy (Anglesey)
Llechfaen (Powys)
Llechylched (Anglesey)
Llidiadnenog (Carmarthenshire)
Llidiart-y-parc (Denbighshire)
Lloc (Flintshire)
LLONG (Flintshire)
Llowes (Powys)
LLWYDCOED (Rhondda Cynon Taff)
Llwydiarth (Powys)
Llwyn (Denbighshire)
Llwyn-drain (Pembrokeshire)
Llwyn-du (Monmouthshire)
Llwyn-on (Merthyr Tydfil)
Llwyn-y-brain (Carmarthenshire)
Llwynderw (Powys)
LLWYNHENDY (Carmarthenshire)
Llwynmawr (Wrexham)
LLWYNYPIA (Rhondda Cynon Taff)
Llyn-y-pandy (Flintshire)
Llynclys (Shropshire)
Llynfaes (Anglesey)
Llys-y-fran (Pembrokeshire)
Llyswen (Powys)
Llysworney (Vale of Glamorgan)
Llywel (Powys)
LOAD BROOK (South Yorkshire)
Loanend (Northumberland)
Locking (Somerset)
Lockington (East Riding of Yorkshire)
Lockington (Leicestershire)
Lockleywood (Shropshire)
Lockton (North Yorkshire)
Loddington (Leicestershire)
Lode Heath (West Midlands)
LODGE GREEN (West Midlands)
LOFHOUSE GATE (West Yorkshire)
LOFTHOUSE (North Yorkshire)
LOFTHOUSE (West Yorkshire)
Loftus (North Yorkshire)
Loganbeck (Cumbria)
Loggerheads (Staffordshire)
Login (Carmarthenshire)
LON-LAS (Swansea)
Londesborough (East Riding of Yorkshire)
London Beach (Kent)
Londonderry (North Yorkshire)
Londonthorpe (Lincolnshire)
LONG ASHTON (Somerset)
LONG BANK (Hereford & Worcester)
Long Bennington (Lincolnshire)
LONG CLAWSON (Leicestershire)
Long Compton (Staffordshire)
Long Compton (Warwickshire)
LONG DRAX (North Yorkshire)
LONG DUCKMANTON (Derbyshire)
Long Eaton (Derbyshire)
Long Green (Cheshire)
Long Green (Hereford & Worcester)
Long Hedges (Lincolnshire)
Long Itchington (Warwickshire)
Long Lane (Shropshire)
Long Lawford (Warwickshire)
Long Load (Somerset)
Long Marston (North Yorkshire)
Long Marston (Warwickshire)
Long Marton (Cumbria)
Long Meadowend (Shropshire)
Long Newnton (Gloucestershire)
Long Preston (North Yorkshire)
Long Riston (East Riding of Yorkshire)
LONG SIGHT (Greater Manchester)
Long Sutton (Lincolnshire)
Long Sutton (Somerset)
Long Waste (Shropshire)
Long Whatton (Leicestershire)
LONGBENTON (Tyne & Wear)
Longborough (Gloucestershire)
Longbridge (Warwickshire)
Longbridge (West Midlands)
Longburgh (Cumbria)
Longcliffe (Derbyshire)
Longcroft (Cumbria)
LONGDEN (Shropshire)
LONGDEN COMMON (Shropshire)
Longdon (Hereford & Worcester)
LONGDON (Staffordshire)
LONGDON GREEN (Staffordshire)
Longdon Heath (Hereford & Worcester)
Longdon upon Tern (Shropshire)
Longfield (Kent)

Longford (Derbyshire)
Longford (Gloucestershire)
Longford (Kent)
LONGFORD (Shropshire)
LONGFORD (West Midlands)
LONGFRAMLINGTON (Northumberland)
LONGHIRST (Northumberland)
Longhope (Gloucestershire)
LONGHORSLEY (Northumberland)
LONGHOUGHTON (Northumberland)
Longlands (Cumbria)
Longlane (Derbyshire)
Longlevens (Gloucestershire)
LONGLEY (West Yorkshire)
Longley Green (Hereford & Worcester)
Longmoss (Cheshire)
Longnewton (Durham)
Longney (Gloucestershire)
LONGNOR (Shropshire)
Longnor (Staffordshire)
Longpark (Cumbria)
Longridge (Lancashire)
Longridge (Staffordshire)
Longsdon (Staffordshire)
LONGSHAW COMMON (Greater Manchester)
Longslow (Shropshire)
LONGSTONE (Pembrokeshire)
Longthwaite (Cumbria)
Longton (Lancashire)
LONGTON (Staffordshire)
Longtown (Cumbria)
Longtown (Hereford & Worcester)
Longville in the Dale (Shropshire)
LONGWITTON (Northumberland)
LONGWOOD (Shropshire)
Loose (Kent)
Loosegate (Lincolnshire)
Lopen (Somerset)
Loppington (Shropshire)
Lorbottle (Northumberland)
Lords Wood (Kent)
LOSCOE (Derbyshire)
Lostford (Shropshire)
Lostock Gralam (Cheshire)
Lostock Green (Cheshire)
LOSTOCK HALL FOLD (Greater Manchester)
LOSTOCK JUNCTION (Greater Manchester)
Lothersdale (North Yorkshire)
Loughborough (Leicestershire)
LOUGHOR (Swansea)
Loughton (Shropshire)
Lound (Lincolnshire)
LOUND (Nottinghamshire)
LOUNT (Leicestershire)
Louth (Lincolnshire)
LOVE CLOUGH (Lancashire)
LOVERSALL (South Yorkshire)
Lovesome Hill (North Yorkshire)
LOVESTON (Pembrokeshire)
Lovington (Somerset)
LOW ACKWORTH (West Yorkshire)
LOW ANGERTON (Northumberland)
Low Barlings (Lincolnshire)
LOW BELL END (North Yorkshire)
LOW BENTHAM (North Yorkshire)
LOW BIGGINS (Cumbria)
Low Borrowbridge (Cumbria)
LOW BRADFIELD (South Yorkshire)
LOW BRADLEY (North Yorkshire)
LOW BRAITHWAITE (Cumbria)
Low Burnham (Lincolnshire)
LOW BUSTON (Northumberland)
Low Catton (East Riding of Yorkshire)
Low Coniscliffe (Durham)
Low Crosby (Cumbria)
Low Dinsdale (Durham)
LOW EGGBOROUGH (North Yorkshire)
Low Ellington (North Yorkshire)
LOW FELL (Tyne & Wear)
Low Gate (Northumberland)
Low Gettbridge (Cumbria)
LOW GRANTLEY (North Yorkshire)
LOW GREEN (North Yorkshire)
LOW HABBERLEY (Hereford & Worcester)
Low Ham (Somerset)
LOW HARROGATE (North Yorkshire)
Low Hawsker (North Yorkshire)
Low Hesket (Cumbria)
Low Hill (Hereford & Worcester)
Low Hutton (North Yorkshire)
Low Knipe (Cumbria)

LOW LAITHE (North Yorkshire)
Low Langton (Lincolnshire)
LOW LEIGHTON (Derbyshire)
Low Lorton (Cumbria)
Low Marnham (Nottinghamshire)
Low Middleton (Northumberland)
LOW MILL (North Yorkshire
Low Moor (Lancashire)
LOW MOOR (West Yorkshire)
LOW MOORSLEY (Tyne & Wear)
Low Mowthorpe (North Yorkshire)
Low Newton (Cumbria)
Low Rogerscales (Cumbria)
LOW ROW (Cumbria)
Low Row (North Yorkshire)
Low Santon (Lincolnshire)
Low Toynton (Lincolnshire)
LOW VALLEY (South Yorkshire)
Low Walworth (Durham)
Low Wood (Cumbria)
Low Worsall (North Yorkshire)
Low Wray (Cumbria)
Lowbands (Gloucestershire)
LOWCA (Cumbria)
LOWDHAM (Nottinghamshire)
Lowe (Shropshire)
Lowe Hill (Staffordshire)
Lower Aisholt (Somerset)
Lower Apperley (Gloucestershire)
Lower Ballam (Lancashire)
Lower Barewood (Hereford & Worcester)
Lower Bartle (Lancashire)
LOWER BAYSTON (Shropshire)
Lower Bentley (Hereford & Worcester)
Lower Beobridge (Shropshire)
LOWER BERRYHILL (Gloucestershire)
LOWER BIRCHWOOD (Derbyshire)
Lower Brailes (Warwickshire)
LOWER BREDBURY (Greater Manchester)
Lower Broadheath (Hereford & Worcester)
Lower Buckenhill (Hereford & Worcester)
Lower Bullingham (Hereford & Worcester)
Lower Burton (Hereford & Worcester)
Lower Cam (Gloucestershire)
Lower Canada (Somerset)
Lower Chapel (Powys)
Lower Clent (Hereford & Worcester)
Lower Clopton (Warwickshire)
Lower Crossings (Derbyshire)
LOWER CUMBERWORTH (West Yorkshire)
LOWER CWMTWRCH (Powys)
LOWER DARWEN (Lancashire)
LOWER DENBY (West Yorkshire)
Lower Dinchope (Shropshire)
Lower Down (Shropshire)
Lower Dunsforth (North Yorkshire)
Lower Egleton (Hereford & Worcester)
Lower Elkstone (Staffordshire)
Lower Ellastone (Staffordshire)
LOWER EYTHORNE (Kent)
Lower Failand (Somerset)
Lower Frankton (Shropshire)
LOWER FREYSTROP (Pembrokeshire)
Lower Godney (Somerset)
LOWER GORNAL (West Midlands)
LOWER GREEN (Greater Manchester)
Lower Green (Kent)
Lower Green (Staffordshire)
Lower Halstow (Kent)
Lower Hardres (Kent)
Lower Harpton (Hereford & Worcester)
LOWER HARTSHAY (Derbyshire)
LOWER HATTON (Staffordshire)
Lower Hawthwaite (Cumbria)
Lower Hergest (Hereford & Worcester)
Lower Heysham (Lancashire)
Lower Higham (Kent)
Lower Hordley (Shropshire)
Lower Howsell (Hereford & Worcester)
Lower Irlam (Greater Manchester)
LOWER KILBURN (Derbyshire)
Lower Kilcott (Gloucestershire)
Lower Kinnerton (Cheshire)
Lower Langford (Somerset)
Lower Leigh (Staffordshire)
Lower Lemington (Gloucestershire)
Lower Llanfadog (Powys)
LOWER LYDBROOK (Gloucestershire)
Lower Lye (Hereford & Worcester)
Lower Machen (Newport)
Lower Maes-coed (Hereford & Worcester)

Lower Marston (Somerset)
Lower Meend (Gloucestershire)
Lower Milton (Somerset)
Lower Moor (West Midlands)
Lower Morton (Gloucestershire)
Lower Norton (Warwickshire)
Lower Penarth (Vale of Glamorgan)
LOWER PENN (Staffordshire)
Lower Penwortham (Lancashire)
Lower Peover (Cheshire)
LOWER PLACE (Greater Manchester)
Lower Quinton (Warwickshire)
Lower Rainham (Kent)
Lower Roadwater (Somerset)
LOWER SALTER (Lancashire)
Lower Shuckburgh (Warwickshire)
Lower Slaughter (Gloucestershire)
LOWER SOOTHILL (West Yorkshire)
LOWER SOUDLEY (Gloucestershire)
Lower Standen (Kent)
Lower Stoke (Kent)
Lower Stone (Gloucestershire)
LOWER STONNALL (Staffordshire)
Lower Stretton (Cheshire)
Lower Swell (Gloucestershire)
LOWER TEAN (Staffordshire)
Lower Town (Hereford & Worcester)
Lower Town (Pembrokeshire)
Lower Tysoe (Warwickshire)
Lower Upnor (Kent)
Lower Vexford (Somerset)
Lower Walton (Cheshire)
Lower Weare (Somerset)
Lower Welson (Hereford & Worcester)
Lower Westmancote (Hereford & Worcester)
Lower Whatley (Somerset)
Lower Whitley (Cheshire)
Lower Wick (Gloucestershire)
Lower Wick (Hereford & Worcester)
Lower Wyche (Hereford & Worcester)
LOWER WYKE (West Yorkshire)
LOWERHOUSE (Lancashire)
Lowesby (Leicestershire)
Loweswater (Cumbria)
Lowgill (Cumbria)
LOWGILL (Lancashire)
Lowick (Cumbria)
LOWICK (Northumberland)
Lowick Bridge (Cumbria)
Lowick Green (Cumbria)
LOWLANDS (Durham)
Lowlands (Torfaen)
Lowsonford (Warwickshire)
LOWTHER (Cumbria)
LOWTHER CASTLE (Cumbria)
Lowthorpe (East Riding of Yorkshire)
LOWTON (Greater Manchester)
Lowton (Somerset)
LOWTON COMMON (Greater Manchester)
LOWTON ST. MARY'S (Greater Manchester)
Loxley (Warwickshire)
Loxley Green (Staffordshire)
Loxter (Hereford & Worcester)
Loxton (Somerset)
Lubenham (Leicestershire)
Lucasgate (Lincolnshire)
Luccombe (Somerset)
Lucker (Northumberland)
Luckwell Bridge (Somerset)
Lucott (Somerset)
Lucton (Hereford & Worcester)
Lucy Cross (North Yorkshire)
Ludborough (Lincolnshire)
Ludchurch (Pembrokeshire)
LUDDENDEN (West Yorkshire)
Luddenden Foot (West Yorkshire)
Luddenham Court (Kent)
Luddesdown (Kent)
Luddington (Lincolnshire)
Luddington (Warwickshire)
Ludford (Lincolnshire)
Ludford (Shropshire)
Ludlow (Shropshire)
Ludney (Lincolnshire)
LUDWORTH (Durham)
Lugg Green (Hereford & Worcester)
LUGSDALE (Cheshire)
Lugwardine (Hereford & Worcester)
Lulham (Hereford & Worcester)
LULLINGTON (Derbyshire)
Lullington (Somerset)

Lulsgate Bottom (Somerset)
Lulsley (Hereford & Worcester)
LUMB (Lancashire)
Lumb (West Yorkshire)
Lumbutts (West Yorkshire)
LUMBY (North Yorkshire)
Lund (East Riding of Yorkshire)
LUND (North Yorkshire)
Lundford Magna (Lincolnshire)
Lunsford (Kent)
Lunt (Merseyside)
Luntley (Hereford & Worcester)
LUPSET (West Yorkshire)
Lupton (Cumbria)
Lusby (Lincolnshire)
Luston (Hereford & Worcester)
LUTLEY (West Midlands)
Luton (Kent)
Lutterworth (Leicestershire)
Lutton (Lincolnshire)
Luxborough (Somerset)
LUZLEY (Greater Manchester)
Lydbury North (Shropshire)
Lydd (Kent)
LYDDEN (Kent)
Lydeard St. Lawrence (Somerset)
Lydford on Fosse (Somerset)
LYDGATE (Greater Manchester)
LYDGATE (West Yorkshire)
Lydham (Shropshire)
Lydiate (Merseyside)
Lydiate Ash (Hereford & Worcester)
LYDNEY (Gloucestershire)
Lydstep (Pembrokeshire)
LYE (West Midlands)
Lye Cross (Somerset)
Lye Green (Warwickshire)
LYE HEAD (Hereford & Worcester)
Lymbridge Green (Kent)
Lyminge (Kent)
Lymm (Cheshire)
Lympne (Kent)
Lympsham (Somerset)
Lynch (Somerset)
Lyndon Green (West Midlands)
Lyne Down (Hereford & Worcester)
Lyneal (Shropshire)
LYNEHOLMFORD (Cumbria)
LYNEMOUTH (Northumberland)
Lyng (Somerset)
Lynhales (Hereford & Worcester)
Lynn (Shropshire)
Lynn (Staffordshire)
Lynsted (Kent)
Lyonshall (Hereford & Worcester)
Lytham (Lancashire)
Lytham St. Anne's (Lancashire)
LYTHBANK (Shropshire)
LYTHE (North Yorkshire)

M

Mablethorpe (Lincolnshire)
MACCLESFIELD (Cheshire)
Macclesfield Forest (Cheshire)
MACHEN (Caerphilly)
Machynlleth (Powys)
MACHYNYS (Carmarthenshire)
Mackworth (Derbyshire)
MADELEY (Shropshire)
MADELEY (Staffordshire)
MADELEY HEATH (Staffordshire)
Madley (Hereford & Worcester)
Madresfield (Hereford & Worcester)
Maenaddwyn (Anglesey)
Maenclochog (Pembrokeshire)
Maendy (Vale of Glamorgan)
Maer (Staffordshire)
MAERDY (Rhondda Cynon Taff)
Maes-glas (Newport)
Maesbrook (Shropshire)
MAESBURY (Shropshire)
Maesbury Marsh (Shropshire)
Maesgwynne (Carmarthenshire)
Maeshafn (Denbighshire)
Maesmynis (Powys)
MAESTEG (Bridgend)
MAESYBONT (Carmarthenshire)
MAESYCWMMER (Caerphilly)
Maghull (Merseyside)
Magor (Monmouthshire)

MAIDEN HEAD (Somerset)
MAIDEN LAW (Durham)
Maiden Wells (Pembrokeshire)
Maidenwell (Lincolnshire)
Maidstone (Kent)
Mainclee (Newport)
MAINSFORTH (Durham)
Mainstone (Shropshire)
Maisemore (Gloucestershire)
Major's Green (Hereford & Worcester)
MAKENEY (Derbyshire)
Malcoff (Derbyshire)
Malham (North Yorkshire)
MALLTRAETH (Anglesey)
Malmsmead (Somerset)
Malpas (Cheshire)
Malpas (Newport)
Maltby (Lincolnshire)
Maltby (North Yorkshire)
MALTBY (South Yorkshire)
Maltby le Marsh (Lincolnshire)
Maltman's Hill (Kent)
Malton (North Yorkshire)
Malvern Link (Hereford & Worcester)
Malvern Wells (Hereford & Worcester)
MAMBLE (Hereford & Worcester)
Mamhilad (Monmouthshire)
Manafon (Powys)
Manby (Lincolnshire)
MANCETTER (Warwickshire)
MANCHESTER (Greater Manchester)
MANCOT (Flintshire)
Maney (West Midlands)
Manfield (North Yorkshire)
MANGOTSFIELD (Gloucestershire)
Mankinholes (West Yorkshire)
Manley (Cheshire)
MANMOEL (Caerphilly)
MANNINGHAM (West Yorkshire)
Manorbier (Pembrokeshire)
Manorbier Newton (Pembrokeshire)
Manordeilo (Carmarthenshire)
Manorowen (Pembrokeshire)
Mansell Gamage (Hereford & Worcester)
Mansell Lacy (Hereford & Worcester)
MANSERGH (Cumbria)
MANSFIELD (Nottinghamshire)
MANSFIELD WOODHOUSE (Nottinghamshire)
Mansriggs (Cumbria)
Manston (Kent)
MANSTON (West Yorkshire)
Manthorpe (Lincolnshire)
Manton (Lincolnshire)
MANTON (Nottinghamshire)
Maperton (Somerset)
MAPLEBECK (Nottinghamshire)
Maplescombe (Kent)
Mapleton (Derbyshire)
Mapleton (Kent)
MAPPERLEY (Derbyshire)
MAPPERLEY PARK (Nottinghamshire)
Mappleborough Green (Warwickshire)
Mappleton (East Riding of Yorkshire)
MAPPLEWELL (South Yorkshire)
Marbury (Cheshire)
Marchamley (Shropshire)
Marchamley Wood (Shropshire)
Marchington (Staffordshire)
Marchington Woodlands (Staffordshire)
MARCHWIEL (Wrexham)
Marcross (Vale of Glamorgan)
Marden (Hereford & Worcester)
Marden (Kent)
Marden Beech (Kent)
Marden Thorn (Kent)
Mardy (Monmouthshire)
Marefield (Leicestershire)
Mareham le Fen (Lincolnshire)
Mareham on the Hill (Lincolnshire)
MAREHAY (Derbyshire)
Marfleet (East Riding of Yorkshire)
MARFORD (Wrexham)
MARGAM (Neath Port Talbot)
Margate (Kent)
Margrove Park (North Yorkshire)
Marian-glas (Anglesey)
Marine Town (Kent)
Mark (Somerset)
Mark Causeway (Somerset)
Markbeech (Kent)
Markby (Lincolnshire)

Markeaton (Derbyshire)
Market Bosworth (Leicestershire)
Market Deeping (Lincolnshire)
Market Drayton (Shropshire)
Market Harborough (Leicestershire)
Market Rasen (Lincolnshire)
Market Stainton (Lincolnshire)
Market Weighton (East Riding of Yorkshire)
Markfield (Leicestershire)
MARKHAM (Caerphilly)
MARKHAM MOOR (Nottinghamshire)
Markington (North Yorkshire)
Marksbury (Somerset)
Marl Bank (Hereford & Worcester)
Marlbrook (Hereford & Worcester)
Marlcliff (Warwickshire)
MARLEY (Kent)
Marley Green (Cheshire)
MARLEY HILL (Tyne & Wear)
Marloes (Pembrokeshire)
Marlow (Hereford & Worcester)
Marlpit Hill (Kent)
MARLPOOL (Derbyshire)
MARPLE (Greater Manchester)
MARPLE BRIDGE (Greater Manchester)
MARR (South Yorkshire)
MARRICK (North Yorkshire)
Marros (Carmarthenshire)
MARSDEN (Tyne & Wear)
Marsden (West Yorkshire)
MARSDEN HEIGHT (Lancashire)
MARSETT (North Yorkshire)
MARSH (West Yorkshire)
Marsh Green (Kent)
Marsh Green (Shropshire)
MARSH GREEN (Staffordshire)
MARSH LANE (Derbyshire)
MARSH LANE (Gloucestershire)
Marsh Street (Somerset)
Marsh The (Powys)
MARSHBOROUGH (Kent)
Marshbrook (Shropshire)
Marshchapel (Lincolnshire)
Marshfield (Gloucestershire)
Marshfield (Newport)
MARSHLAND GREEN (Greater Manchester)
Marshside (Merseyside)
MARSKE (North Yorkshire)
Marske-by-the-Sea (North Yorkshire)
Marston (Cheshire)
Marston (Hereford & Worcester)
Marston (Lincolnshire)
Marston (Staffordshire)
MARSTON (Warwickshire)
Marston Green (West Midlands)
MARSTON JABBET (Warwickshire)
Marston Magna (Somerset)
Marston Montgomery (Derbyshire)
Marston on Dove (Derbyshire)
Marston Stannet (Hereford & Worcester)
Marstow (Hereford & Worcester)
Marthall (Cheshire)
MARTIN (Kent)
Martin (Lincolnshire)
Martin Dales (Lincolnshire)
Martin Hussingtree (Hereford & Worcester)
Martindale (Cumbria)
MARTINSCROFT (Cheshire)
MARTLETWY (Pembrokeshire)
Martley (Hereford & Worcester)
Martock (Somerset)
Marton (Cheshire)
Marton (East Riding of Yorkshire)
Marton (Lincolnshire)
Marton (North Yorkshire)
Marton (Shropshire)
Marton (Warwickshire)
Marton-le-Moor (North Yorkshire)
Maryland (Monmouthshire)
MARYLEBONE (Greater Manchester)
MARYPORT (Cumbria)
Masham (North Yorkshire)
MASON (Tyne & Wear)
MASONGILL (North Yorkshire)
MASTIN MOOR (Derbyshire)
MATFEN (Northumberland)
Matfield (Kent)
Mathern (Monmouthshire)
Mathon (Hereford & Worcester)
Mathry (Pembrokeshire)
Matlock (Derbyshire)

Matlock Bank (Derbyshire)
Matlock Bath (Derbyshire)
Matlock Dale (Derbyshire)
Matson (Gloucestershire)
Matterdale End (Cumbria)
MATTERSEY (Nottinghamshire)
MATTERSEY THORPE (Nottinghamshire)
Maugersbury (Gloucestershire)
MAULDS MEABURN (Cumbria)
Maunby (North Yorkshire)
Maund Bryan (Hereford & Worcester)
Maundown (Somerset)
MAVESYN RIDWARE (Staffordshire)
Mavis Enderby (Lincolnshire)
Maw Green (Cheshire)
MAW GREEN (West Midlands)
Mawbray (Cumbria)
MAWDESLEY (Lancashire)
Mawdlam (Bridgend)
Mawthorpe (Lincolnshire)
MAXSTOKE (Warwickshire)
Maxted Street (Kent)
Maxton (Kent)
MAY BANK (Staffordshire)
MAYALS (Swansea)
Mayfield (Staffordshire)
Maypole (Kent)
Maypole (Monmouthshire)
Maypole (West Midlands)
MEADGATE (Somerset)
MEADOWFIELD (Durham)
MEADOWHALL (South Yorkshire)
Meadowtown (Shropshire)
Meal Bank (Cumbria)
MEALRIGG (Cumbria)
MEALSGATE (Cumbria)
MEANWOOD (West Yorkshire)
Mearbeck (North Yorkshire)
Meare (Somerset)
Meare Green (Somerset)
MEASHAM (Leicestershire)
Meathop (Cumbria)
Meaux (East Riding of Yorkshire)
Medbourne (Leicestershire)
MEDEN VALE (Nottinghamshire)
Medlam (Lincolnshire)
Medlar (Lancashire)
MEDOMSLEY (Durham)
Meer Common (Hereford & Worcester)
Meerbrook (Staffordshire)
Meeson (Shropshire)
Meidrim (Carmarthenshire)
Meifod (Powys)
Meinciau (Carmarthenshire)
MEIR (Staffordshire)
MEIR HEATH (Staffordshire)
Melbourne (Derbyshire)
Melbourne (East Riding of Yorkshire)
MELDON (Northumberland)
Meldon Park (Northumberland)
Meliden (Denbighshire)
MELIN COURT (Neath Port Talbot)
Melin-byrhedyn (Powys)
Melin-y-ddol (Powys)
Melin-y-wig (Denbighshire)
MELINAU (Pembrokeshire)
MELKINTHORPE (Cumbria)
MELKRIDGE (Northumberland)
Mellguards (Cumbria)
MELLING (Lancashire)
Melling (Merseyside)
Melling Mount (Merseyside)
MELLOR (Greater Manchester)
MELLOR (Lancashire)
MELLOR BROOK (Lancashire)
MELLS (Somerset)
Melmerby (Cumbria)
MELMERBY (North Yorkshire)
Melsonby (North Yorkshire)
MELTHAM (West Yorkshire)
MELTHAM MILLS (West Yorkshire)
Melton (East Riding of Yorkshire)
MELTON MOWBRAY (Leicestershire)
Melton Ross (Lincolnshire)
Meltonby (East Riding of Yorkshire)
Melverley (Shropshire)
Melverley Green (Shropshire)
Menai Bridge (Anglesey)
MENITHWOOD (Hereford & Worcester)
MENSTON (West Yorkshire)
MENTHORPE (North Yorkshire)

MEOLE BRACE (Shropshire)
Meopham (Kent)
Meopham Green (Kent)
Meopham Station (Kent)
Mere (Cheshire)
Mere Brow (Lancashire)
Mere Green (Hereford & Worcester)
Mere Green (West Midlands)
Mere Heath (Cheshire)
MERECLOUGH (Lancashire)
Meresborough (Kent)
Mereworth (Kent)
MERIDEN (West Midlands)
Merlin's Bridge (Pembrokeshire)
Merrington (Shropshire)
Merrion (Pembrokeshire)
Merriott (Somerset)
MERRY HILL (West Midlands)
MERRY LEES (Leicestershire)
MERRYHILL (West Midlands)
Mersham (Kent)
Merthyr (Carmarthenshire)
Merthyr Cynog (Powys)
Merthyr Dyfan (Vale of Glamorgan)
Merthyr Mawr (Bridgend)
MERTHYR TYDFIL (Merthyr Tydfil)
MERTHYR VALE (Merthyr Tydfil)
Messingham (Lincolnshire)
Metheringham (Lincolnshire)
METHLEY (West Yorkshire)
METHLEY JUNCTION (West Yorkshire)
MEXBOROUGH (South Yorkshire)
Meysey Hampton (Gloucestershire)
Michaelchurch (Hereford & Worcester)
Michaelchurch Escley (Hereford & Worcester)
Michaelchurch-on-Arrow (Powys)
Michaelston-le-Pit (Vale of Glamorgan)
Michaelstone-y-Fedw (Newport)
MICKLE TRAFFORD (Cheshire)
MICKLEBRING (South Yorkshire)
MICKLEBY (North Yorkshire)
MICKLEFIELD (West Yorkshire)
Mickleover (Derbyshire)
Micklethwaite (Cumbria)
MICKLETHWAITE (West Yorkshire)
Mickleton (Durham)
Mickleton (Gloucestershire)
MICKLETOWN (West Yorkshire)
MICKLEY (Derbyshire)
Mickley (North Yorkshire)
MICKLEY SQUARE (Northumberland)
Mid Thorpe (Lincolnshire)
Middle Chinnock (Somerset)
Middle Duntisbourne (Gloucestershire)
MIDDLE HANDLEY (Derbyshire)
Middle Littleton (Hereford & Worcester)
MIDDLE MADELEY (Staffordshire)
Middle Maes-coed (Hereford & Worcester)
Middle Mayfield (Staffordshire)
Middle Mill (Pembrokeshire)
Middle Quarter (Kent)
Middle Rasen (Lincolnshire)
MIDDLE SALTER (Lancashire)
Middle Stoford (Somerset)
Middle Stoke (Kent)
Middle Stoughton (Somerset)
Middle Street (Gloucestershire)
Middle Tysoe (Warwickshire)
Middle Yard (Gloucestershire)
MIDDLECLIFFE (South Yorkshire)
Middleham (North Yorkshire)
Middlehope (Shropshire)
Middlesbrough (North Yorkshire)
MIDDLESCEUGH (Cumbria)
Middleshaw (Cumbria)
MIDDLESMOOR (North Yorkshire)
MIDDLESTONE (Durham)
MIDDLESTONE MOOR (Durham)
MIDDLESTOWN (West Yorkshire)
Middleton (Cumbria)
Middleton (Derbyshire)
Middleton (Durham)
MIDDLETON (Greater Manchester)
Middleton (Hereford & Worcester)
Middleton (Lancashire)
Middleton (North Yorkshire)
Middleton (Northumberland)
Middleton (Shropshire)
Middleton (Swansea)
Middleton (Warwickshire)
MIDDLETON (West Yorkshire)

Middleton Green (Staffordshire)
Middleton Hall (Northumberland)
Middleton on the Hill (Hereford & Worcester)
Middleton One Row (Durham)
Middleton Priors (Shropshire)
Middleton Quernhow (North Yorkshire)
MIDDLETON SCRIVEN (Shropshire)
Middleton St. George (Durham)
Middleton Tyas (North Yorkshire)
Middleton-in-Teesdale (Durham)
Middleton-on-Leven (North Yorkshire)
Middleton-on-the-Wolds (East Riding of Yorkshire)
Middletown (Cumbria)
Middletown (Powys)
Middletown (Somerset)
Middlewich (Cheshire)
Middlewood (Hereford & Worcester)
Middlezoy (Somerset)
MIDDRIDGE (Durham)
Midford (Somerset)
Midge Hall (Lancashire)
MIDGEHOLME (Cumbria)
MIDGLEY (West Yorkshire)
MIDHOPESTONES (South Yorkshire)
Midney (Somerset)
MIDSOMER NORTON (Somerset)
Midville (Lincolnshire)
MIDWAY (Cheshire)
Milborne Port (Somerset)
Milborne Wick (Somerset)
Milbourne (Northumberland)
Milburn (Cumbria)
Milbury Heath (Gloucestershire)
Milby (North Yorkshire)
Mile End (Gloucestershire)
Mile Oak (Kent)
MILE OAK (Staffordshire)
Mile Town (Kent)
Milebrook (Powys)
Milebush (Kent)
Miles Hope (Hereford & Worcester)
MILES PLATTING (Greater Manchester)
Milfield (Northumberland)
Milford (Derbyshire)
Milford (Powys)
MILFORD (Staffordshire)
Milford Haven (Pembrokeshire)
MILKWALL (Gloucestershire)
Mill Bank (West Yorkshire)
MILL BROW (Greater Manchester)
Mill Green (Lincolnshire)
Mill Green (Staffordshire)
MILL GREEN (West Midlands)
Mill Meece (Staffordshire)
Mill Side (Cumbria)
Mill Street (Kent)
Millbeck (Cumbria)
MILLBROOK (Greater Manchester)
Milldale (Staffordshire)
MILLEND (Gloucestershire)
Miller's Dale (Derbyshire)
Millers Green (Derbyshire)
MILLGATE (Lancashire)
Millgreen (Shropshire)
Millhalf (Hereford & Worcester)
Millhead (Lancashire)
MILLHOUSE (Cumbria)
MILLHOUSE GREEN (South Yorkshire)
MILLHOUSES (South Yorkshire)
MILLIN CROSS (Pembrokeshire)
Millington (East Riding of Yorkshire)
Millness (Cumbria)
Millom (Cumbria)
MILLTHORPE (Derbyshire)
Millthrop (Cumbria)
MILLTOWN (Derbyshire)
MILNROW (Greater Manchester)
Milnthorpe (Cumbria)
MILNTHORPE (West Yorkshire)
Milson (Shropshire)
Milsted (Kent)
Milthorpe (Lincolnshire)
MILTON (Cumbria)
MILTON (Derbyshire)
Milton (Kent)
Milton (Newport)
MILTON (Nottinghamshire)
Milton (Pembrokeshire)
Milton (Somerset)
MILTON (Staffordshire)

See paras. 2 and 4 of the User Guide 2003 if you can't find your place name.

Milton Clevedon (Somerset)
Milton End (Gloucestershire)
Milton Green (Cheshire)
Milton Regis (Kent)
Milverton (Somerset)
Milverton (Warwickshire)
Milwich (Staffordshire)
Milwr (Flintshire)
Minchinhampton (Gloucestershire)
Mindrum (Northumberland)
Mindrum Mill (Northumberland)
Minehead (Somerset)
MINERA (Wrexham)
Miningsby (Lincolnshire)
Minnis Bay (Kent)
Minskip (North Yorkshire)
Minster (Kent)
MINSTERACRES (Northumberland)
MINSTERLEY (Shropshire)
Minsterworth (Gloucestershire)
Minting (Lincolnshire)
Minton (Shropshire)
MINWEAR (Pembrokeshire)
MINWORTH (West Midlands)
MIREHOUSE (Cumbria)
MIRFIELD (West Yorkshire)
Miserden (Gloucestershire)
MISKIN (Rhondda Cynon Taff)
MISSON (Nottinghamshire)
Misterton (Leicestershire)
MISTERTON (Nottinghamshire)
Misterton (Somerset)
Mitchel Troy (Monmouthshire)
MITCHELDEAN (Gloucestershire)
MITFORD (Northumberland)
Mitton (Staffordshire)
Mixenden (West Yorkshire)
Mixon (Staffordshire)
Mobberley (Cheshire)
MOBBERLEY (Staffordshire)
Moccas (Hereford & Worcester)
Mochdre (Powys)
Mockbeggar (Kent)
MOCKERKIN (Cumbria)
MODDERSHALL (Staffordshire)
Moelfre (Anglesey)
Moelfre (Powys)
MOIRA (Leicestershire)
Molash (Kent)
MOLD (Flintshire)
MOLDGREEN (West Yorkshire)
Molescroft (East Riding of Yorkshire)
MOLESDEN (Northumberland)
Mollington (Cheshire)
Monday Boys (Kent)
Monington (Pembrokeshire)
MONK BRETTON (South Yorkshire)
MONK FRYSTON (North Yorkshire)
MONK HESLEDEN (Durham)
Monkhide (Hereford & Worcester)
Monkhill (Cumbria)
Monkhopton (Shropshire)
Monkland (Hereford & Worcester)
Monknash (Vale of Glamorgan)
Monks Heath (Cheshire)
Monks Horton (Kent)
Monks Kirby (Warwickshire)
MONKSEATON (Tyne & Wear)
Monksilver (Somerset)
Monkspath (West Midlands)
Monksthorpe (Lincolnshire)
Monkswood (Monmouthshire)
Monkton (Kent)
MONKTON (Tyne & Wear)
Monkton (Vale of Glamorgan)
Monkton Combe (Somerset)
Monkton Heathfield (Somerset)
MONKWEARMOUTH (Tyne & Wear)
MONMORE GREEN (West Midlands)
Monmouth (Monmouthshire)
Monnington on Wye (Hereford & Worcester)
Montacute (Somerset)
MONTCLIFFE (Greater Manchester)
Montford (Shropshire)
Montford Bridge (Shropshire)
Montgomery (Powys)
MONTON (Greater Manchester)
MONTPELIER (Bristol)
Monyash (Derbyshire)
MOOR ALLERTON (West Yorkshire)
Moor End (East Riding of Yorkshire)

Moor End (Lancashire)
MOOR END (North Yorkshire)
MOOR END (West Yorkshire)
MOOR HEAD (West Yorkshire)
Moor Monkton (North Yorkshire)
MOOR ROW (Cumbria)
Moor Row (Durham)
Moor Side (Lancashire)
Moor Side (Lincolnshire)
Moor Street (Kent)
MOOR STREET (West Midlands)
Moorby (Lincolnshire)
Moorcot (Hereford & Worcester)
Moore (Cheshire)
Moorend (Gloucestershire)
MOORENDS (South Yorkshire)
MOORGREEN (Nottinghamshire)
MOORHALL (Derbyshire)
Moorhampton (Hereford & Worcester)
Moorhouse (Cumbria)
MOORHOUSE (Nottinghamshire)
MOORHOUSE (West Yorkshire)
Moorland (Somerset)
Moorlinch (Somerset)
Moorsholm (North Yorkshire)
Moorside (Cumbria)
MOORSIDE (Greater Manchester)
Moorstock (Kent)
MOORTHORPE (West Yorkshire)
Moortown (Lincolnshire)
Moortown (Shropshire)
MOORTOWN (West Yorkshire)
MORDA (Shropshire)
Mordiford (Hereford & Worcester)
Mordon (Durham)
More (Shropshire)
Morecambe (Lancashire)
Morehall (Kent)
MORESBY (Cumbria)
MORESBY PARKS (Cumbria)
Moreton (Hereford & Worcester)
Moreton (Merseyside)
Moreton (Staffordshire)
Moreton Corbet (Shropshire)
Moreton Jeffries (Hereford & Worcester)
Moreton Mill (Shropshire)
Moreton Morrell (Warwickshire)
Moreton on Lugg (Hereford & Worcester)
Moreton Paddox (Warwickshire)
Moreton Say (Shropshire)
Moreton Valence (Gloucestershire)
Moreton-in-Marsh (Gloucestershire)
MORFA GLAS (Neath Port Talbot)
MORGANSTOWN (Cardiff)
Morland (Cumbria)
Morley (Cheshire)
MORLEY (Derbyshire)
MORLEY (Durham)
MORLEY (West Yorkshire)
Morley Green (Cheshire)
MORPETH (Northumberland)
Morrey (Staffordshire)
Morridge Side (Staffordshire)
MORRIDGE TOP (Staffordshire)
MORRISTON (Swansea)
MORTHEN (South Yorkshire)
Mortimer's Cross (Hereford & Worcester)
Morton (Cumbria)
MORTON (Derbyshire)
Morton (Lincolnshire)
Morton (Nottinghamshire)
Morton (Shropshire)
Morton Hall (Lincolnshire)
Morton Tinmouth (Durham)
Morton-on-Swale (North Yorkshire)
MORVILLE (Shropshire)
MORVILLE HEATH (Shropshire)
MOSBOROUGH (South Yorkshire)
Mose (Shropshire)
Mosedale (Cumbria)
Moseley (Hereford & Worcester)
Moseley (West Midlands)
MOSES GATE (Greater Manchester)
MOSS (South Yorkshire)
MOSS (Wrexham)
MOSS BANK (Merseyside)
Moss Edge (Lancashire)
Moss End (Cheshire)
Moss Side (Cumbria)
Moss Side (Lancashire)
Moss Side (Merseyside)

MOSSBAY (Cumbria)
Mossbrow (Greater Manchester)
Mosser Mains (Cumbria)
Mossley (Cheshire)
MOSSLEY (Greater Manchester)
Mossley Hill (Merseyside)
MOSSY LEA (Lancashire)
MOSTON (Greater Manchester)
Moston (Shropshire)
Moston Green (Cheshire)
MOSTYN (Flintshire)
Motherby (Cumbria)
MOTTRAM IN LONGDENDALE (Greater Manchester)
Mottram St Andrew (Cheshire)
Mouldsworth (Cheshire)
Moulton (Cheshire)
Moulton (Lincolnshire)
Moulton (North Yorkshire)
Moulton (Vale of Glamorgan)
Moulton Chapel (Lincolnshire)
Moulton Seas End (Lincolnshire)
MOUNT (West Yorkshire)
MOUNT PLEASANT (Cheshire)
MOUNT PLEASANT (Derbyshire)
MOUNT PLEASANT (Durham)
Mount Pleasant (Hereford & Worcester)
MOUNT TABOR (West Yorkshire)
MOUNTAIN (West Yorkshire)
MOUNTAIN ASH (Rhondda Cynon Taff)
Mountain Street (Kent)
Mounton (Monmouthshire)
Mountsorrel (Leicestershire)
MOW COP (Cheshire)
Mowmacre Hill (Leicestershire)
Mowsley (Leicestershire)
Moylgrove (Pembrokeshire)
Much Birch (Hereford & Worcester)
Much Cowarne (Hereford & Worcester)
Much Dewchurch (Hereford & Worcester)
Much Hoole (Lancashire)
Much Hoole Town (Lancashire)
Much Marcle (Hereford & Worcester)
Much Wenlock (Shropshire)
Muchelney (Somerset)
Muchelney Ham (Somerset)
Mucklestone (Staffordshire)
Muckley (Shropshire)
Muckton (Lincolnshire)
Mud Row (Kent)
Mudford (Somerset)
Mudford Sock (Somerset)
Mudgley (Somerset)
Mugginton (Derbyshire)
Muggintonlane End (Derbyshire)
Muggleswick (Durham)
Muker (North Yorkshire)
Mumby (Lincolnshire)
Munderfield Row (Hereford & Worcester)
Munderfield Stocks (Hereford & Worcester)
Mungrisdale (Cumbria)
Munsley (Hereford & Worcester)
Munslow (Shropshire)
Murcot (Hereford & Worcester)
Murston (Kent)
MURTON (Cumbria)
MURTON (Durham)
Murton (North Yorkshire)
MURTON (Northumberland)
MURTON (Tyne & Wear)
Muscoates (North Yorkshire)
MUSTON (Leicestershire)
Muston (North Yorkshire)
Mustow Green (Hereford & Worcester)
MUXTON (Shropshire)
Myddfai (Carmarthenshire)
Myddle (Shropshire)
Mynachlog ddu (Pembrokeshire)
Myndd-llan (Flintshire)
Myndtown (Shropshire)
MYNYDD ISA (Flintshire)
Mynydd Mechell (Anglesey)
Mynydd-bach (Monmouthshire)
MYNYDD-BACH (Swansea)
MYNYDDGARREG (Carmarthenshire)
Mytholm (West Yorkshire)
MYTHOLMROYD (West Yorkshire)
Mythop (Lancashire)
Myton-on-Swale (North Yorkshire)

See paras. 2 and 4 of the User Guide 2003 if you can't find your place name.

N

Nab's Head (Lancashire)
NABURN (North Yorkshire)
Nackholt (Kent)
Nackington (Kent)
Nafferton (East Riding of Yorkshire)
Nag's Head (Gloucestershire)
NAILBRIDGE (Gloucestershire)
Nailsbourne (Somerset)
NAILSEA (Somerset)
NAILSTONE (Leicestershire)
Nailsworth (Gloucestershire)
Nannerch (Flintshire)
Nanpantan (Leicestershire)
Nant-ddu (Powys)
Nant-glas (Powys)
NANT-Y-BWCH (Blaenau Gwent)
Nant-y-caws (Carmarthenshire)
Nant-y-derry (Monmouthshire)
NANT-Y-GOLLEN (Shropshire)
NANT-Y-MOEL (Bridgend)
Nantgaredig (Carmarthenshire)
NANTGARW (Rhondda Cynon Taff)
Nantglyn (Denbighshire)
Nantgwyn (Powys)
Nantmawr (Shropshire)
Nantmel (Powys)
Nantwich (Cheshire)
NANTYFFYLLON (Bridgend)
Napleton (Hereford & Worcester)
Nappa (North Yorkshire)
Napton on the Hill (Warwickshire)
Narberth (Pembrokeshire)
Narborough (Leicestershire)
Nash (Hereford & Worcester)
Nash (Newport)
Nash (Shropshire)
NASH END (Hereford & Worcester)
Nash Street (Kent)
Nastend (Gloucestershire)
NATEBY (Cumbria)
Nateby (Lancashire)
Natland (Cumbria)
Naunton (Gloucestershire)
Naunton (Hereford & Worcester)
Naunton Beauchamp (Hereford & Worcester)
Navenby (Lincolnshire)
Nawton (North Yorkshire)
Neal's Green (Warwickshire)
Near Cotton (Staffordshire)
Near Sawry (Cumbria)
Neasham (Durham)
NEATH (Neath Port Talbot)
Nebo (Anglesey)
NEDDERTON (Northumberland)
NEEN SAVAGE (Shropshire)
NEEN SOLLARS (Shropshire)
Neenton (Shropshire)
NELSON (Caerphilly)
NELSON (Lancashire)
Nempnett Thrubwell (Somerset)
NENTHALL (Cumbria)
NENTHEAD (Cumbria)
NERCWYS (Flintshire)
Nesbit (Northumberland)
NESFIELD (North Yorkshire)
NESS (Cheshire)
Nesscliffe (Shropshire)
NESTON (Cheshire)
Netchwood (Shropshire)
Nether Alderley (Cheshire)
NETHER BROUGHTON (Nottinghamshire)
NETHER BURROW (Lancashire)
NETHER HANDLEY (Derbyshire)
NETHER HAUGH (South Yorkshire)
Nether Headon (Nottinghamshire)
NETHER HEAGE (Derbyshire)
NETHER KELLET (Lancashire)
NETHER LANGWITH (Nottinghamshire)
NETHER MOOR (Derbyshire)
Nether Padley (Derbyshire)
Nether Poppleton (North Yorkshire)
Nether Row (Cumbria)
Nether Silton (North Yorkshire)
Nether Skyborry (Shropshire)
Nether Stowey (Somerset)
Nether Wasdale (Cumbria)
NETHER WELTON (Cumbria)
Nether Westcote (Gloucestershire)
NETHER WHITACRE (Warwickshire)

Netherby (Cumbria)
Netherby (North Yorkshire)
Nethercote (Warwickshire)
Netherend (Gloucestershire)
Netherfield (Leicestershire)
NETHERFIELD (Nottinghamshire)
Nethergate (Lincolnshire)
Netherland Green (Staffordshire)
NETHEROYD HILL (West Yorkshire)
NETHERSEAL (Derbyshire)
NETHERTHONG (West Yorkshire)
NETHERTHORPE (Derbyshire)
Netherton (Hereford & Worcester)
Netherton (Merseyside)
NETHERTON (Northumberland)
NETHERTON (Shropshire)
NETHERTON (West Midlands)
NETHERTON (West Yorkshire)
Nethertown (Cumbria)
Nethertown (Lancashire)
NETHERTOWN (Staffordshire)
NETHERWITTON (Northumberland)
NETTLEBRIDGE (Somerset)
Nettleham (Lincolnshire)
Nettlestead (Kent)
Nettlestead Green (Kent)
NETTLESWORTH (Durham)
Nettleton (Lincolnshire)
Neuadd (Carmarthenshire)
Neuadd Fawr (Carmarthenshire)
Neuadd-ddu (Powys)
Nevern (Pembrokeshire)
Nevill Holt (Leicestershire)
New Arram (East Riding of Yorkshire)
New Ash Green (Kent)
New Balderton (Nottinghamshire)
New Barn (Kent)
New Bewick (Northumberland)
New Bilton (Warwickshire)
New Bolingbroke (Lincolnshire)
New Boultham (Lincolnshire)
NEW BRAMPTON (Derbyshire)
NEW BRANCEPETH (Durham)
New Bridge (North Yorkshire)
NEW BRIGHTON (Flintshire)
New Brighton (Merseyside)
NEW BRINSLEY (Nottinghamshire)
New Brotton (North Yorkshire)
NEW BROUGHTON (Wrexham)
NEW BURY (Greater Manchester)
NEW CLIPSTONE (Nottinghamshire)
NEW COWPER (Cumbria)
NEW CROFTON (West Yorkshire)
New Cross (Somerset)
NEW DELAVAL (Northumberland)
NEW DELPH (Greater Manchester)
New Earswick (North Yorkshire)
NEW EASTWOOD (Nottinghamshire)
NEW EDLINGTON (South Yorkshire)
New Ellerby (East Riding of Yorkshire)
New End (Hereford & Worcester)
NEW FARNLEY (West Yorkshire)
New Ferry (Merseyside)
NEW FRYSTON (West Yorkshire)
NEW HARTLEY (Northumberland)
NEW HEDGES (Pembrokeshire)
NEW HERRINGTON (Tyne & Wear)
New Holland (Lincolnshire)
NEW HOUGHTON (Derbyshire)
NEW HOUSES (Greater Manchester)
New Houses (North Yorkshire)
New Hutton (Cumbria)
New Hythe (Kent)
New Inn (Carmarthenshire)
New Inn (Torfaen)
New Invention (Shropshire)
NEW LAMBTON (Durham)
New Lane (Lancashire)
NEW LANE END (Cheshire)
New Leake (Lincolnshire)
New Longton (Lancashire)
New Marton (Shropshire)
NEW MILL (West Yorkshire)
NEW MILLS (Derbyshire)
New Mills (Powys)
New Moat (Pembrokeshire)
NEW OLLERTON (Nottinghamshire)
NEW OSCOTT (West Midlands)
New Radnor (Powys)
New Rent (Cumbria)
NEW RIDLEY (Northumberland)

New Road Side (North Yorkshire)
New Romney (Kent)
NEW ROSSINGTON (South Yorkshire)
New Row (Lancashire)
NEW SHARLSTON (West Yorkshire)
NEW SHORESTON (Northumberland)
NEW SILKSWORTH (Tyne & Wear)
New Skelton (North Yorkshire)
New Somerby (Lincolnshire)
New Spilsby (Lincolnshire)
NEW SPRINGS (Greater Manchester)
New Street (Hereford & Worcester)
NEW SWANNINGTON (Leicestershire)
New Town (Somerset)
NEW TREDEGAR (Caerphilly)
NEW TUPTON (Derbyshire)
New Village (East Riding of Yorkshire)
New Waltham (Lincolnshire)
NEW WHITTINGTON (Derbyshire)
New York (Lincolnshire)
NEW YORK (North Yorkshire)
NEW YORK (Tyne & Wear)
New Zealand (Derbyshire)
Newall (West Yorkshire)
Newark-on-Trent (Nottinghamshire)
Newbarn (Kent)
NEWBIGGIN (Cumbria)
NEWBIGGIN (Durham)
NEWBIGGIN (North Yorkshire)
NEWBIGGIN-BY-THE-SEA (Northumberland)
NEWBIGGIN-ON-LUNE (Cumbria)
NEWBOLD (Derbyshire)
NEWBOLD (Leicestershire)
Newbold on Avon (Warwickshire)
Newbold on Stour (Warwickshire)
Newbold Pacey (Warwickshire)
Newbold Revel (Warwickshire)
NEWBOLD VERDON (Leicestershire)
Newborough (Anglesey)
Newborough (Staffordshire)
NEWBOTTLE (Tyne & Wear)
NEWBRIDGE (Caerphilly)
Newbridge (Pembrokeshire)
NEWBRIDGE (Wrexham)
Newbridge Green (Hereford & Worcester)
Newbridge on Wye (Powys)
Newbridge-on-Usk (Monmouthshire)
NEWBROUGH (Northumberland)
NEWBURGH (Lancashire)
Newburgh Priory (North Yorkshire)
NEWBURN (Tyne & Wear)
NEWBURY (Somerset)
NEWBY (Cumbria)
Newby (Lancashire)
NEWBY (North Yorkshire)
Newby Bridge (Cumbria)
Newby Cross (Cumbria)
Newby East (Cumbria)
NEWBY HEAD (Cumbria)
Newby West (Cumbria)
Newby Wiske (North Yorkshire)
Newcastle (Monmouthshire)
Newcastle (Shropshire)
Newcastle Emlyn (Carmarthenshire)
NEWCASTLE UPON TYNE (Tyne & Wear)
NEWCASTLE-UNDER-LYME (Staffordshire)
Newchapel (Pembrokeshire)
NEWCHAPEL (Staffordshire)
NEWCHURCH (Blaenau Gwent)
Newchurch (Hereford & Worcester)
Newchurch (Kent)
Newchurch (Powys)
Newchurch (Staffordshire)
NEWCHURCH IN PENDLE (Lancashire)
Newenden (Kent)
NEWENT (Gloucestershire)
NEWFIELD (Durham)
NEWGALE (Pembrokeshire)
Newhall (Cheshire)
NEWHALL (Derbyshire)
Newham (Northumberland)
Newhaven (Derbyshire)
NEWHEY (Greater Manchester)
NEWHOLM (North Yorkshire)
Newingreen (Kent)
Newington (Kent)
Newington (Shropshire)
Newington Bagpath (Gloucestershire)
Newland (Cumbria)
Newland (East Riding of Yorkshire)
NEWLAND (Gloucestershire)

See paras. 2 and 4 of the User Guide 2003 if you can't find your place name.

Newland (Hereford & Worcester)
NEWLAND (North Yorkshire)
Newland (Somerset)
NEWLANDS (Cumbria)
NEWLANDS (Northumberland)
NEWMARKET (Cumbria)
NEWMILLERDAM (West Yorkshire)
Newmills (Monmouthshire)
Newnes (Shropshire)
Newnham (Gloucestershire)
Newnham (Hereford & Worcester)
Newnham (Kent)
Newnham Paddox (Warwickshire)
Newport (East Riding of Yorkshire)
Newport (Gloucestershire)
Newport (Newport)
Newport (Pembrokeshire)
NEWPORT (Shropshire)
NEWSAM GREEN (West Yorkshire)
Newsbank (Cheshire)
Newsham (Lancashire)
NEWSHAM (North Yorkshire)
NEWSHAM (Northumberland)
Newsholme (East Riding of Yorkshire)
Newsholme (Lancashire)
NEWSTEAD (Northumberland)
NEWSTEAD (Nottinghamshire)
NEWTHORPE (North Yorkshire)
Newton (Cardiff)
Newton (Cheshire)
Newton (Cumbria)
NEWTON (Derbyshire)
Newton (Hereford & Worcester)
NEWTON (Lancashire)
Newton (Lincolnshire)
Newton (North Yorkshire)
NEWTON (Northumberland)
NEWTON (Nottinghamshire)
Newton (Shropshire)
Newton (Somerset)
Newton (Staffordshire)
Newton (Warwickshire)
NEWTON (West Midlands)
NEWTON (West Yorkshire)
Newton Arlosh (Cumbria)
NEWTON AYCLIFFE (Durham)
Newton Bewley (Durham)
Newton Burgoland (Leicestershire)
Newton by Toft (Lincolnshire)
Newton Green (Monmouthshire)
Newton Harcourt (Leicestershire)
NEWTON HEATH (Greater Manchester)
NEWTON HILL (West Yorkshire)
Newton Kyme (North Yorkshire)
Newton Morrell (North Yorkshire)
Newton Mountain (Pembrokeshire)
NEWTON MULGRAVE (North Yorkshire)
Newton on Ouse (North Yorkshire)
Newton on the Hill (Shropshire)
NEWTON ON TRENT (Lincolnshire)
NEWTON REGIS (Warwickshire)
Newton Reigny (Cumbria)
NEWTON SOLNEY (Derbyshire)
NEWTON ST. LOE (Somerset)
Newton under Roseberry (North Yorkshire)
NEWTON UNDERWOOD (Northumberland)
Newton upon Derwent (East Riding of Yorkshire)
Newton-by-the-Sea (Northumberland)
NEWTON-LE-WILLOWS (Merseyside)
Newton-le-Willows (North Yorkshire)
NEWTON-ON-THE-MOOR (Northumberland)
NEWTOWN (Bristol)
Newtown (Cheshire)
Newtown (Cumbria)
NEWTOWN (Derbyshire)
Newtown (Gloucestershire)
NEWTOWN (Greater Manchester)
Newtown (Hereford & Worcester)
NEWTOWN (Lancashire)
Newtown (Northumberland)
Newtown (Powys)
NEWTOWN (Rhondda Cynon Taff)
Newtown (Shropshire)
NEWTOWN (Staffordshire)
Newtown Linford (Leicestershire)
Newtown Unthank (Leicestershire)
Nextend (Hereford & Worcester)
Neyland (Pembrokeshire)
NIBLEY (Gloucestershire)
Nibley Green (Gloucestershire)
Nicholaston (Swansea)

Nickies Hill (Cumbria)
Nidd (North Yorkshire)
Nine Wells (Pembrokeshire)
NINEBANKS (Northumberland)
Nineveh (Hereford & Worcester)
Niwbwrch (Anglesey)
No Man's Heath (Cheshire)
NO MAN'S HEATH (Warwickshire)
Noah's Ark (Kent)
NOBLETHORPE (West Yorkshire)
NOBOLD (Shropshire)
Nocton (Lincolnshire)
Nolton (Pembrokeshire)
NOLTON HAVEN (Pembrokeshire)
Noneley (Shropshire)
NONINGTON (Kent)
NOOK (Cumbria)
Norbreck (Lancashire)
Norbridge (Hereford & Worcester)
Norbury (Cheshire)
Norbury (Derbyshire)
Norbury (Shropshire)
Norbury (Staffordshire)
Norbury Common (Cheshire)
Norbury Junction (Staffordshire)
Norchard (Hereford & Worcester)
Norcott Brook (Cheshire)
Norcross (Lancashire)
NORDEN (Greater Manchester)
NORDLEY (Shropshire)
Norham (Northumberland)
Norland Town (West Yorkshire)
Norley (Cheshire)
Normanby (Lincolnshire)
Normanby (North Yorkshire)
Normanby le Wold (Lincolnshire)
Normanton (Derbyshire)
Normanton (Leicestershire)
Normanton (Lincolnshire)
Normanton (Nottinghamshire)
NORMANTON (West Yorkshire)
NORMANTON LE HEATH (Leicestershire)
Normanton on Soar (Nottinghamshire)
NORMANTON ON THE WOLDS (Nottinghamshire)
NORMANTON ON TRENT (Nottinghamshire)
Normoss (Lancashire)
Norris Green (Merseyside)
NORRISTHORPE (West Yorkshire)
NORTH ANSTON (South Yorkshire)
North Barrow (Somerset)
NORTH BIDDICK (Tyne & Wear)
NORTH BITCHBURN (Durham)
North Brewham (Somerset)
North Cadbury (Somerset)
North Carlton (Lincolnshire)
NORTH CARLTON (Nottinghamshire)
North Cave (East Riding of Yorkshire)
North Cerney (Gloucestershire)
NORTH CHARLTON (Northumberland)
North Cheriton (Somerset)
North Cliffe (East Riding of Yorkshire)
NORTH CLIFTON (Nottinghamshire)
NORTH CLOSE (Durham)
North Cockerington (Lincolnshire)
North Collingham (Nottinghamshire)
North Cornelly (Bridgend)
North Cotes (Lincolnshire)
North Cowton (North Yorkshire)
North Curry (Somerset)
North Dalton (East Riding of Yorkshire)
North Deighton (North Yorkshire)
NORTH DUFFIELD (North Yorkshire)
North Elham (Kent)
North Elkington (Lincolnshire)
NORTH ELMSALL (West Yorkshire)
North End (Cumbria)
North End (East Riding of Yorkshire)
North End (Leicestershire)
North End (Lincolnshire)
North End (Merseyside)
North End (Northumberland)
North End (Somerset)
North Evington (Leicestershire)
North Ferriby (East Riding of Yorkshire)
North Frodingham (East Riding of Yorkshire)
North Grimston (North Yorkshire)
North Halling (Kent)
NORTH HAZELRIGG (Northumberland)
North Hele (Somerset)
North Hykeham (Lincolnshire)
North Kelsey (Lincolnshire)

North Killingholme (Lincolnshire)
North Kilvington (North Yorkshire)
North Kilworth (Leicestershire)
North Kyme (Lincolnshire)
North Landing (East Riding of Yorkshire)
North Lees (North Yorkshire)
North Leigh (Kent)
North Leverton with Habblesthorpe (Nottinghamshire)
North Littleton (Hereford & Worcester)
North Middleton (Northumberland)
NORTH MUSKHAM (Nottinghamshire)
North Newbald (East Riding of Yorkshire)
North Newton (Somerset)
North Nibley (Gloucestershire)
North Ormsby (Lincolnshire)
North Otterington (North Yorkshire)
North Owersby (Lincolnshire)
North Perrott (Somerset)
North Petherton (Somerset)
North Piddle (Hereford & Worcester)
North Quarme (Somerset)
North Rauceby (Lincolnshire)
North Reston (Lincolnshire)
North Rigton (North Yorkshire)
North Rode (Cheshire)
North Row (Cumbria)
North Scale (Cumbria)
NORTH SCARLE (Lincolnshire)
NORTH SEATON (Northumberland)
NORTH SEATON COLLIERY (Northumberland)
NORTH SHIELDS (Tyne & Wear)
North Shore (Lancashire)
NORTH SIDE (Cumbria)
North Skelton (North Yorkshire)
North Somercotes (Lincolnshire)
North Stainley (North Yorkshire)
NORTH STAINMORE (Cumbria)
NORTH STOKE (Somerset)
North Street (Kent)
NORTH SUNDERLAND (Northumberland)
North Thoresby (Lincolnshire)
North Town (Somerset)
NORTH WALBOTTLE (Tyne & Wear)
NORTH WHEATLEY (Nottinghamshire)
North Wick (Somerset)
NORTH WIDCOMBE (Somerset)
North Willingham (Lincolnshire)
NORTH WINGFIELD (Derbyshire)
North Witham (Lincolnshire)
North Wootton (Somerset)
Northallerton (North Yorkshire)
Northampton (Hereford & Worcester)
Northay (Somerset)
NORTHBOURNE (Kent)
Northdown (Kent)
NORTHEDGE (Derbyshire)
Northend (Warwickshire)
NORTHENDEN (Greater Manchester)
Northfield (East Riding of Yorkshire)
Northfield (West Midlands)
Northfields (Lincolnshire)
Northfleet (Kent)
Northington (Gloucestershire)
Northlands (Lincolnshire)
Northleach (Gloucestershire)
Northload Bridge (Somerset)
Northmoor (Somerset)
NORTHOP (Flintshire)
NORTHOP HALL (Flintshire)
Northorpe (Lincolnshire)
NORTHORPE (West Yorkshire)
Northover (Somerset)
NORTHOWRAM (West Yorkshire)
NORTHVILLE (Bristol)
Northway (Somerset)
NORTHWAY (Swansea)
Northwich (Cheshire)
Northwick (Gloucestershire)
Northwick (Hereford & Worcester)
Northwick (Somerset)
Northwood (Derbyshire)
Northwood (Shropshire)
NORTHWOOD (Staffordshire)
Northwood Green (Gloucestershire)
Norton (Cheshire)
Norton (Durham)
Norton (Gloucestershire)
Norton (Hereford & Worcester)
Norton (Monmouthshire)
Norton (North Yorkshire)

See paras. 2 and 4 of the User Guide 2003 if you can't find your place name.

NORTON (Nottinghamshire)
Norton (Powys)
NORTON (Shropshire)
Norton (Somerset)
NORTON (South Yorkshire)
NORTON (Swansea)
Norton Bridge (Staffordshire)
NORTON CANES (Staffordshire)
Norton Canon (Hereford & Worcester)
NORTON DISNEY (Lincolnshire)
Norton Fitzwarren (Somerset)
NORTON GREEN (Staffordshire)
Norton Hawkfield (Somerset)
Norton in Hales (Shropshire)
NORTON IN THE MOORS (Staffordshire)
Norton Lindsey (Warwickshire)
NORTON MALREWARD (Somerset)
Norton St. Philip (Somerset)
Norton sub Hamdon (Somerset)
Norton Wood (Hereford & Worcester)
Norton-Juxta-Twycross (Leicestershire)
Norton-le-Clay (North Yorkshire)
NORWELL (Nottinghamshire)
NORWELL WOODHOUSE (Nottinghamshire)
Norwood (Kent)
NORWOOD (South Yorkshire)
NORWOOD GREEN (West Yorkshire)
Noseley (Leicestershire)
Nosterfield (North Yorkshire)
Notgrove (Gloucestershire)
Nottage (Bridgend)
NOTTINGHAM (Nottinghamshire)
NOTTON (West Yorkshire)
Nottswood Hill (Gloucestershire)
Noutard's Green (Hereford & Worcester)
NOX (Shropshire)
Nun Monkton (North Yorkshire)
Nunburnholme (East Riding of Yorkshire)
NUNCARGATE (Nottinghamshire)
Nunclose (Cumbria)
NUNEATON (Warwickshire)
Nunkeeling (East Riding of Yorkshire)
Nunney (Somerset)
Nunney Catch (Somerset)
Nunnington (Hereford & Worcester)
Nunnington (North Yorkshire)
NUNNYKIRK (Northumberland)
NUNS MOOR (Tyne & Wear)
Nunsthorpe (Lincolnshire)
Nunthorpe (North Yorkshire)
Nunthorpe Village (North Yorkshire)
Nunwick (North Yorkshire)
Nunwick (Northumberland)
Nupdown (Gloucestershire)
Nupend (Gloucestershire)
Nurton (Staffordshire)
NUTHALL (Nottinghamshire)
NUTTAL LANE (Greater Manchester)
NUTWELL (South Yorkshire)
Nympsfield (Gloucestershire)
Nynehead (Somerset)
Nythe (Somerset)

O

Oad Street (Kent)
Oadby (Leicestershire)
Oak Tree (Durham)
Oakall Green (Hereford & Worcester)
OAKAMOOR (Staffordshire)
OAKDALE (Caerphilly)
Oake (Somerset)
Oaken (Staffordshire)
Oakenclough (Lancashire)
OAKENGATES (Shropshire)
OAKENHOLT (Flintshire)
OAKENSHAW (Durham)
OAKENSHAW (West Yorkshire)
Oaker Side (Derbyshire)
OAKERTHORPE (Derbyshire)
Oakgrove (Cheshire)
Oakhanger (Cheshire)
OAKHILL (Somerset)
Oakhurst (Kent)
Oaklands (Powys)
Oakle Street (Gloucestershire)
Oakley Park (Powys)
Oakridge (Gloucestershire)
Oaks (Lancashire)
OAKS (Shropshire)
Oaks Green (Derbyshire)

OAKSHAW (Cumbria)
OAKTHORPE (Leicestershire)
Oakwood (Derbyshire)
OAKWOOD (Northumberland)
OAKWORTH (West Yorkshire)
Oare (Kent)
Oare (Somerset)
Oasby (Lincolnshire)
Oath (Somerset)
Obley (Shropshire)
Obthorpe (Lincolnshire)
Occlestone Green (Cheshire)
Ockbrook (Derbyshire)
OCKER HILL (West Midlands)
Ockeridge (Hereford & Worcester)
Ocle Pychard (Hereford & Worcester)
Octon (East Riding of Yorkshire)
Odcombe (Somerset)
Odd Down (Somerset)
Oddingley (Hereford & Worcester)
Oddington (Gloucestershire)
ODSAL (West Yorkshire)
ODSTONE (Leicestershire)
Offchurch (Warwickshire)
Offenham (Hereford & Worcester)
OFFERTON (Tyne & Wear)
Offham (Kent)
Offleymarsh (Shropshire)
OGDEN (West Yorkshire)
Ogle (Northumberland)
Oglet (Merseyside)
Ogmore (Vale of Glamorgan)
OGMORE VALE (Bridgend)
Ogmore-by-Sea (Vale of Glamorgan)
OLD BASFORD (Nottinghamshire)
Old Bewick (Northumberland)
Old Bolingbroke (Lincolnshire)
Old Bramhope (West Yorkshire)
OLD BRAMPTON (Derbyshire)
OLD BYLAND (North Yorkshire)
OLD CASSOP (Durham)
Old Castle (Bridgend)
Old Church Stoke (Powys)
Old Clee (Lincolnshire)
Old Cleeve (Somerset)
OLD CLIPSTONE (Nottinghamshire)
OLD DALBY (Leicestershire)
Old Dam (Derbyshire)
Old Ditch (Somerset)
OLD EDINGTON (South Yorkshire)
OLD ELDON (Durham)
Old Ellerby (East Riding of Yorkshire)
Old Forge (Hereford & Worcester)
Old Furnace (Hereford & Worcester)
OLD GLOSSOP (Derbyshire)
OLD GOOLE (East Riding of Yorkshire)
OLD HILL (West Midlands)
Old Hutton (Cumbria)
OLD LANGHO (Lancashire)
Old Leake (Lincolnshire)
Old Malton (North Yorkshire)
OLD MICKLEFIELD (West Yorkshire)
Old Milverton (Warwickshire)
OLD QUARRINGTON (Durham)
OLD RADFORD (Nottinghamshire)
Old Radnor (Powys)
Old Romney (Kent)
Old Soar (Kent)
Old Sodbury (Gloucestershire)
Old Somerby (Lincolnshire)
OLD SUNNFORD (West Midlands)
Old Swan (Merseyside)
OLD SWINFORD (West Midlands)
Old Tebay (Cumbria)
Old Thirsk (North Yorkshire)
OLD TOWN (Cumbria)
Old Town (Northumberland)
OLD TOWN (West Yorkshire)
OLD TRAFFORD (Greater Manchester)
OLD TUPTON (Derbyshire)
Old Wives Lees (Kent)
Oldberrow (Warwickshire)
Oldbury (Kent)
OLDBURY (Shropshire)
OLDBURY (Warwickshire)
OLDBURY (West Midlands)
Oldbury Naite (Gloucestershire)
Oldbury on the Hill (Gloucestershire)
Oldbury-on-Severn (Gloucestershire)
Oldcastle (Monmouthshire)
Oldcastle Heath (Cheshire)

OLDCOTES (Nottinghamshire)
Oldfield (Hereford & Worcester)
OLDFIELD (West Yorkshire)
Oldford (Somerset)
OLDHAM (Greater Manchester)
OLDLAND (Gloucestershire)
Oldmixon (Somerset)
OLDSTEAD (North Yorkshire)
Oldwall (Cumbria)
Oldwalls (Swansea)
Oldwoods (Shropshire)
OLIVE GREEN (Staffordshire)
Ollerton (Cheshire)
OLLERTON (Nottinghamshire)
Ollerton (Shropshire)
Olton (West Midlands)
Olveston (Gloucestershire)
Ombersley (Hereford & Worcester)
OMPTON (Nottinghamshire)
Once Brewed (Northumberland)
Onecote (Staffordshire)
Onen (Monmouthshire)
Ongar Street (Hereford & Worcester)
Onibury (Shropshire)
ONLLWYN (Neath Port Talbot)
Onneley (Staffordshire)
Onston (Cheshire)
OPENWOODGATE (Derbyshire)
Orby (Lincolnshire)
Orchard Portman (Somerset)
Orcop (Hereford & Worcester)
Orcop Hill (Hereford & Worcester)
Ordley (Northumberland)
ORDSALL (Nottinghamshire)
Oreleton Common (Hereford & Worcester)
Oreton (Shropshire)
ORFORD (Cheshire)
Orgreave (Staffordshire)
Orlestone (Kent)
Orleton (Hereford & Worcester)
Ormathwaite (Cumbria)
Ormesby (North Yorkshire)
ORMSKIRK (Lancashire)
ORNSBY HILL (Durham)
ORRELL (Greater Manchester)
Orrell (Merseyside)
ORRELL POST (Greater Manchester)
Orslow (Staffordshire)
Orston (Nottinghamshire)
Orthwaite (Cumbria)
Ortner (Lancashire)
Orton (Cumbria)
ORTON (Staffordshire)
Orton Rigg (Cumbria)
Orton-on-the-Hill (Leicestershire)
Osbaldeston (Lancashire)
Osbaldeston Green (Lancashire)
Osbaldwick (North Yorkshire)
OSBASTON (Leicestershire)
Osbaston (Shropshire)
Osbournby (Lincolnshire)
Oscroft (Cheshire)
OSGATHORPE (Leicestershire)
Osgodby (Lincolnshire)
OSGODBY (North Yorkshire)
OSMANTHORPE (West Yorkshire)
Osmaston (Derbyshire)
OSMONDTHORPE (West Yorkshire)
Osmotherley (North Yorkshire)
Ospringe (Kent)
OSSETT (West Yorkshire)
OSSINGTON (Nottinghamshire)
Oswaldkirk (North Yorkshire)
OSWALDTWISTLE (Lancashire)
OSWESTRY (Shropshire)
Otford (Kent)
Otham (Kent)
Otham Hole (Kent)
Othery (Somerset)
Otley (West Yorkshire)
Otterburn (North Yorkshire)
OTTERBURN (Northumberland)
Otterham Quay (Kent)
Otterhampton (Somerset)
Otterspool (Merseyside)
Ottinge (Kent)
Ottringham (East Riding of Yorkshire)
Oughterby (Cumbria)
Oughtershaw (North Yorkshire)
OUGHTERSIDE (Cumbria)
OUGHTIBRIDGE (South Yorkshire)

See paras. 2 and 4 of the User Guide 2003 if you can't find your place name.

59

Oughtrington (Cheshire)
Oulston (North Yorkshire)
Oulton (Cumbria)
Oulton (Staffordshire)
OULTON (West Yorkshire)
OUNSDALE (Staffordshire)
Ousby (Cumbria)
Ousefleet (East Riding of Yorkshire)
OUSTON (Durham)
Ouston (Northumberland)
OUT ELMSTEAD (Kent)
Out Newton (East Riding of Yorkshire)
Out Rawcliffe (Lancashire)
OUTCHESTER (Northumberland)
Outgate (Cumbria)
OUTHGILL (Cumbria)
Outhill (Warwickshire)
Outlands (Staffordshire)
OUTLANE (West Yorkshire)
OUTWOOD (West Yorkshire)
OUTWOOD GATE (Greater Manchester)
OUTWOODS (Leicestershire)
Outwoods (Staffordshire)
OUTWOODS (Warwickshire)
OUZLEWELL GREEN (West Yorkshire)
OVENDEN (West Yorkshire)
Over (Cheshire)
OVER (Gloucestershire)
Over Burrows (Derbyshire)
Over Green (Warwickshire)
Over Haddon (Derbyshire)
OVER KELLET (Lancashire)
Over Monnow (Monmouthshire)
Over Silton (North Yorkshire)
Over Stowey (Somerset)
Over Stratton (Somerset)
Over Tabley (Cheshire)
OVER WHITACRE (Warwickshire)
OVER WOODHOUSE (Derbyshire)
Overbury (Hereford & Worcester)
OVERGREEN (Derbyshire)
Overleigh (Somerset)
Overley (Staffordshire)
Overpool (Cheshire)
OVERSEAL (Derbyshire)
Oversland (Kent)
Overstey Green (Warwickshire)
Overton (Cheshire)
Overton (Lancashire)
Overton (North Yorkshire)
Overton (Shropshire)
Overton (Swansea)
OVERTON (West Yorkshire)
OVERTON (Wrexham)
OVERTON BRIDGE (Wrexham)
Overton Green (Cheshire)
OVERTOWN (Lancashire)
OVERTOWN (West Yorkshire)
OVINGHAM (Northumberland)
OVINGTON (Durham)
OVINGTON (Northumberland)
Owlbury (Shropshire)
OWLERTON (South Yorkshire)
Owlpen (Gloucestershire)
Owmby (Lincolnshire)
Owston (Leicestershire)
OWSTON (South Yorkshire)
Owston Ferry (Lincolnshire)
Owstwick (East Riding of Yorkshire)
Owthorne (East Riding of Yorkshire)
OWTHORPE (Nottinghamshire)
Oxcombe (Lincolnshire)
OXCROFT (Derbyshire)
Oxen Park (Cumbria)
Oxenholme (Cumbria)
OXENHOPE (West Yorkshire)
Oxenpill (Somerset)
Oxenton (Gloucestershire)
OXHILL (Durham)
Oxhill (Warwickshire)
OXLEY (West Midlands)
OXSPRING (South Yorkshire)
Oxton (North Yorkshire)
OXTON (Nottinghamshire)
Oxwich (Swansea)
Oxwich Green (Swansea)
Oystermouth (Swansea)
Ozleworth (Gloucestershire)

P

PACKINGTON (Leicestershire)
PACKMOOR (Staffordshire)
Packmores (Warwickshire)
PADDINGTON (Cheshire)
Paddington (Merseyside)
Paddlesworth (Kent)
Paddock Wood (Kent)
Paddolgreen (Shropshire)
PADESWOOD (Flintshire)
Padfield (Derbyshire)
PADGATE (Cheshire)
PADIHAM (Lancashire)
PADSIDE (North Yorkshire)
PAGE BANK (Durham)
Pailton (Warwickshire)
Painleyhill (Staffordshire)
Painscastle (Powys)
PAINSHAWFIELD (Northumberland)
Painsthorpe (East Riding of Yorkshire)
Painswick (Gloucestershire)
Painter's Forstal (Kent)
PALFREY (West Midlands)
Palmerstown (Vale of Glamorgan)
PALTERTON (Derbyshire)
Pamington (Gloucestershire)
Panborough (Somerset)
Pancross (Vale of Glamorgan)
Pandy (Powys)
Pandy (Wrexham)
Pandy'r Capel (Denbighshire)
Panks Bridge (Hereford & Worcester)
Pannal (North Yorkshire)
Pannal Ash (North Yorkshire)
Pant (Shropshire)
Pant Mawr (Powys)
Pant-Gwyn (Carmarthenshire)
PANT-LASAU (Swansea)
Pant-pastynog (Denbighshire)
Pant-y-dwr (Powys)
Pant-y-ffridd (Powys)
PANT-Y-GOG (Bridgend)
Pant-y-mwyn (Flintshire)
Pantasaph (Flintshire)
Panteg (Pembrokeshire)
Pantglas (Powys)
Panton (Lincolnshire)
PANTYFFYNNON (Carmarthenshire)
PANTYGASSEG (Torfaen)
Pantymenyn (Carmarthenshire)
Papcastle (Cumbria)
PAPPLEWICK (Nottinghamshire)
Paramour Street (Kent)
PARBOLD (Lancashire)
Parbrook (Somerset)
Parc Seymour (Newport)
PARDSHAW (Cumbria)
PARK (Northumberland)
PARK BRIDGE (Greater Manchester)
Park End (Northumberland)
PARK END (Staffordshire)
Park Gate (Hereford & Worcester)
PARK GATE (West Yorkshire)
Park Head (Cumbria)
PARK HEAD (Derbyshire)
Park Hill (Gloucestershire)
PARKEND (Gloucestershire)
Parkers Green (Kent)
Parkgate (Cheshire)
PARKGATE (Cumbria)
Parkgate (Kent)
Parkhill (Nottinghamshire)
Parkhouse (Monmouthshire)
Parkmill (Swansea)
PARKSIDE (Durham)
PARKSIDE (Wrexham)
PARLINGTON (West Yorkshire)
PARR BRIDGE (Greater Manchester)
Parrah Green (Cheshire)
Parrog (Pembrokeshire)
PARSON'S CROSS (South Yorkshire)
Parson's Hill (Derbyshire)
PARSONBY (Cumbria)
Partington (Greater Manchester)
Partney (Lincolnshire)
PARTON (Cumbria)
Partrishow (Powys)
Parwich (Derbyshire)
Pasturefields (Staffordshire)
Patchway (Gloucestershire)

PATELEY BRIDGE (North Yorkshire)
Pateshall (Hereford & Worcester)
Pathe (Somerset)
Pathlow (Warwickshire)
Patrick Brompton (North Yorkshire)
PATRICROFT (Greater Manchester)
Patrington (East Riding of Yorkshire)
Patrixbourne (Kent)
Patterdale (Cumbria)
Pattingham (Staffordshire)
Patton (Shropshire)
Paull (East Riding of Yorkshire)
PAULTON (Somerset)
Paunton (Hereford & Worcester)
PAUPERHAUGH (Northumberland)
PAVE LANE (Shropshire)
Pawlett (Somerset)
Pawston (Northumberland)
Paxford (Gloucestershire)
Payden Street (Kent)
Paythorne (Lancashire)
Paytoe (Hereford & Worcester)
Peak Dale (Derbyshire)
Peak Forest (Derbyshire)
Peak Hill (Lincolnshire)
Pearson's Green (Kent)
Peartree Green (Hereford & Worcester)
PEASEDOWN ST. JOHN (Somerset)
PEASEHILL (Derbyshire)
PEASLEY CROSS (Merseyside)
Peasmarsh (Somerset)
Peatling Magna (Leicestershire)
Peatling Parva (Leicestershire)
Peaton (Shropshire)
Pebworth (Hereford & Worcester)
PECKET WELL (West Yorkshire)
Peckforton (Cheshire)
Peckleton (Leicestershire)
Pedair-ffordd (Powys)
Pedlinge (Kent)
PEDMORE (West Midlands)
Pedwell (Somerset)
Peel (Lancashire)
Peene (Kent)
Peening Quarter (Kent)
PEGSWOOD (Northumberland)
Pegwell (Kent)
PELAW (Tyne & Wear)
Pelcomb (Pembrokeshire)
Pelcomb Bridge (Pembrokeshire)
Pelcomb Cross (Pembrokeshire)
PELSALL (West Midlands)
PELSALL WOOD (West Midlands)
PELTON (Durham)
PELTON FELL (Durham)
Pelutho (Cumbria)
PEMBERTON (Carmarthenshire)
PEMBERTON (Greater Manchester)
Pembles Cross (Kent)
PEMBREY (Carmarthenshire)
Pembridge (Hereford & Worcester)
Pembroke (Pembrokeshire)
Pembroke Dock (Pembrokeshire)
Pembury (Kent)
Pen-ffordd (Pembrokeshire)
Pen-groes-oped (Monmouthshire)
Pen-llyn (Anglesey)
Pen-lon (Anglesey)
Pen-rhiw (Pembrokeshire)
PEN-TWYN (Caerphilly)
PEN-TWYN (Torfaen)
Pen-y-bont (Powys)
Pen-y-Bont-Fawr (Powys)
PEN-Y-BRYN (Neath Port Talbot)
Pen-y-bryn (Pembrokeshire)
PEN-Y-CAE (Powys)
Pen-y-cae-mawr (Monmouthshire)
Pen-y-cefn (Flintshire)
Pen-y-clawdd (Monmouthshire)
PEN-Y-COEDCAE (Rhondda Cynon Taff)
PEN-Y-CWN (Pembrokeshire)
PEN-Y-DARREN (Merthyr Tydfil)
PEN-Y-FAI (Bridgend)
Pen-y-felin (Flintshire)
Pen-y-garnedd (Anglesey)
Pen-y-genffordd (Powys)
Pen-y-lan (Vale of Glamorgan)
Pen-y-stryt (Denbighshire)
Pen-yr-Heol (Monmouthshire)
PEN-YR-HEOLGERRIG (Merthyr Tydfil)
Penallt (Monmouthshire)

Penally (Pembrokeshire)
Penalt (Hereford & Worcester)
Penarth (Vale of Glamorgan)
Penblewin (Pembrokeshire)
Pencader (Carmarthenshire)
Pencarnisiog (Anglesey)
Pencarreg (Carmarthenshire)
Pencelli (Powys)
PENCLAWDD (Swansea)
PENCOED (Bridgend)
Pencombe (Hereford & Worcester)
Pencoyd (Hereford & Worcester)
Pencraig (Hereford & Worcester)
Pencraig (Powys)
Penderyn (Rhondda Cynon Taff)
Pendine (Carmarthenshire)
PENDLEBURY (Greater Manchester)
Pendleton (Lancashire)
Pendock (Hereford & Worcester)
Pendomer (Somerset)
Pendoylan (Vale of Glamorgan)
Pendre (Bridgend)
Penegoes (Powys)
PENGAM (Caerphilly)
Pengam (Cardiff)
Pengorffwysfa (Anglesey)
Pengwern (Denbighshire)
Penhow (Newport)
Peniel (Carmarthenshire)
Peniel (Denbighshire)
PENISTONE (South Yorkshire)
PENKETH (Cheshire)
Penkridge (Staffordshire)
PENLEY (Wrexham)
PENLLERGAER (Swansea)
Penllyn (Vale of Glamorgan)
PENMAEN (Caerphilly)
Penmaen (Swansea)
Penmark (Vale of Glamorgan)
Penmon (Anglesey)
Penmynydd (Anglesey)
PENN (West Midlands)
Pennant (Denbighshire)
Pennant (Powys)
Pennant-Melangell (Powys)
Pennard (Swansea)
Pennerley (Shropshire)
Pennington (Cumbria)
PENNINGTON GREEN (Greater Manchester)
Pennorth (Powys)
Pennsylvania (Gloucestershire)
Penny Bridge (Cumbria)
Penny Bridge (Pembrokeshire)
PENNY GREEN (Nottinghamshire)
Penny Hill (Lincolnshire)
PENPEDAIRHEOL (Caerphilly)
Penpedairheol (Monmouthshire)
Penperlleni (Monmouthshire)
Penpont (Powys)
Penrherber (Carmarthenshire)
PENRHIWCEIBER (Rhondda Cynon Taff)
Penrhos (Anglesey)
PENRHOS (Powys)
Penrice (Swansea)
PENRITH (Cumbria)
Penruddock (Cumbria)
Pensarn (Carmarthenshire)
PENSAX (Hereford & Worcester)
Pensby (Merseyside)
Penselwood (Somerset)
PENSFORD (Somerset)
Pensham (Hereford & Worcester)
PENSHAW (Tyne & Wear)
Penshurst (Kent)
Penshurst Station (Kent)
PENSNETT (West Midlands)
Penstrowed (Powys)
PENTLEPOIR (Pembrokeshire)
Pentraeth (Anglesey)
Pentre (Powys)
Pentre (Shropshire)
PENTRE BACH (Flintshire)
PENTRE BERW (Anglesey)
PENTRE FFWRNDAN (Flintshire)
Pentre Halkyn (Flintshire)
Pentre Hodrey (Shropshire)
Pentre Llanrhaeadr (Denbighshire)
Pentre Llifior (Powys)
Pentre Meyrick (Vale of Glamorgan)
Pentre Saron (Denbighshire)
Pentre ty gwyn (Carmarthenshire)

Pentre'r-felin (Powys)
Pentre-bâch (Powys)
Pentre-cagel (Cardiganshire)
Pentre-celyn (Denbighshire)
Pentre-celyn (Powys)
PENTRE-CHWYTH (Swansea)
Pentre-clawdd (Shropshire)
Pentre-cwrt (Carmarthenshire)
PENTRE-DWR (Swansea)
PENTRE-GWENLAIS (Carmarthenshire)
Pentre-llwyn-llwyd (Powys)
Pentre-Maw (Powys)
PENTRE-PIOD (Torfaen)
Pentre-poeth (Newport)
PENTREBACH (Merthyr Tydfil)
Pentrebeirdd (Powys)
Pentredwr (Denbighshire)
Pentrefelin (Anglesey)
Pentregalar (Pembrokeshire)
PENTRICH (Derbyshire)
PENTYRCH (Cardiff)
Penwyllt (Powys)
Penybanc (Carmarthenshire)
Penybont (Powys)
PENYCAE (Wrexham)
PENYFFORDD (Flintshire)
Penygarnedd (Powys)
PENYGRAIG (Rhondda Cynon Taff)
PENYGROES (Carmarthenshire)
Penysarn (Anglesey)
PENYWAUN (Rhondda Cynon Taff)
PENYWERN (Neath Port Talbot)
Peopleton (Hereford & Worcester)
Peover Heath (Cheshire)
Peplow (Shropshire)
Periton (Somerset)
PERKINSVILLE (Durham)
PERLETHORPE (Nottinghamshire)
Perrott's Brook (Gloucestershire)
Perry (West Midlands)
PERRY BARR (West Midlands)
Perry Street (Somerset)
Pershall (Staffordshire)
Pershore (Hereford & Worcester)
Perthy (Shropshire)
PERTON (Hereford & Worcester)
PERTON (Staffordshire)
Pet Street (Kent)
Peterchurch (Hereford & Worcester)
PETERLEE (Durham)
Peterston-Super-Ely (Vale of Glamorgan)
Peterstone Wentlooge (Newport)
Peterstow (Hereford & Worcester)
Petham (Kent)
Pett Bottom (Kent)
Petton (Shropshire)
Petty France (Gloucestershire)
Phepson (Hereford & Worcester)
PHILADELPHIA (Tyne & Wear)
Phocle Green (Hereford & Worcester)
Pibsbury (Somerset)
PICA (Cumbria)
PICCADILLY (Warwickshire)
Pickering (North Yorkshire)
PICKFORD (West Midlands)
PICKFORD GREEN (West Midlands)
Pickhill (North Yorkshire)
Picklescott (Shropshire)
Pickmere (Cheshire)
Pickney (Somerset)
Pickstock (Shropshire)
PICKUP BANK (Lancashire)
Pickwell (Leicestershire)
Pickworth (Lincolnshire)
Pict's Cross (Hereford & Worcester)
Picton (Cheshire)
PICTON (Flintshire)
Picton (North Yorkshire)
Picton Ferry (Carmarthenshire)
Pie Corner (Hereford & Worcester)
Piercebridge (Durham)
Piff's Elm (Gloucestershire)
Pig Street (Hereford & Worcester)
PIGDON (Northumberland)
Pigeon Green (Warwickshire)
Pikehall (Derbyshire)
Pilham (Lincolnshire)
Pill (Somerset)
PILLATON (Staffordshire)
Pillerton Hersey (Warwickshire)
Pillerton Priors (Warwickshire)

Pilleth (Powys)
PILLEY (South Yorkshire)
Pillgwenlly (Newport)
Pilling (Lancashire)
Pilling Lane (Lancashire)
Pilning (Gloucestershire)
Pilot Inn (Kent)
Pilsbury (Derbyshire)
PILSLEY (Derbyshire)
Pilton (Somerset)
Pilton Green (Swansea)
Pimlico (Lancashire)
Pinchbeck (Lincolnshire)
Pinchbeck Bars (Lincolnshire)
Pinchbeck West (Lincolnshire)
PINCHEON GREEN (South Yorkshire)
Pinchinthorpe (North Yorkshire)
PINCOCK (Lancashire)
Pinfold (Lancashire)
PINGED (Carmarthenshire)
PINKETT'S BOOTH (West Midlands)
PINLEY (West Midlands)
Pinley Green (Warwickshire)
Pinsley Green (Cheshire)
Pinvin (Hereford & Worcester)
PINXTON (Derbyshire)
Pipe and Lyde (Hereford & Worcester)
Pipe Gate (Shropshire)
Pipehill (Staffordshire)
PIPPIN STREET (Lancashire)
Pipton (Powys)
Pirton (Hereford & Worcester)
Pitchcombe (Gloucestershire)
Pitcher Row (Lincolnshire)
Pitchford (Shropshire)
Pitcombe (Somerset)
Pitcot (Vale of Glamorgan)
Pitney (Somerset)
PITSES (Greater Manchester)
Pitt Court (Gloucestershire)
Pitt's Wood (Kent)
PITTINGTON (Durham)
PITY ME (Durham)
Pivington (Kent)
Plaish (Shropshire)
PLAISTOW (Derbyshire)
Plaistow (Hereford & Worcester)
PLANK LANE (Greater Manchester)
Plas Cymyran (Anglesey)
Platt (Kent)
PLATT BRIDGE (Greater Manchester)
Platt Lane (Shropshire)
Platts Heath (Kent)
PLAWSWORTH (Durham)
Plaxtol (Kent)
Playley Green (Gloucestershire)
PLEALEY (Shropshire)
PLEASINGTON (Lancashire)
PLEASLEY (Derbyshire)
PLEASLEYHILL (Nottinghamshire)
PLEDWICK (West Yorkshire)
Plemstall (Cheshire)
PLENMELLER (Northumberland)
Ploughfield (Hereford & Worcester)
Plowden (Shropshire)
PLOXGREEN (Shropshire)
Pluckley (Kent)
Pluckley Station (Kent)
Pluckley Thorne (Kent)
Plucks Gutter (Kent)
PLUMBLAND (Cumbria)
Plumgarths (Cumbria)
Plumley (Cheshire)
Plumpton (Cumbria)
Plumpton Head (Cumbria)
PLUMTREE (Nottinghamshire)
Plumtree Green (Kent)
PLUNGAR (Leicestershire)
Plurenden (Kent)
POCKLEY (North Yorkshire)
Pocklington (East Riding of Yorkshire)
Pode Hole (Lincolnshire)
Podimore (Somerset)
Podmore (Staffordshire)
Pointon (Lincolnshire)
Pole Elm (Hereford & Worcester)
Pole Moor (West Yorkshire)
Polelane Ends (Cheshire)
POLESWORTH (Warwickshire)
POLLINGTON (East Riding of Yorkshire)
Polsham (Somerset)

See paras. 2 and 4 of the User Guide 2003 if you can't find your place name.

Ponde (Powys)
Ponsonby (Cumbria)
PONT MORLAIS (Carmarthenshire)
PONT RHYD-Y-CYFF (Bridgend)
Pont Robert (Powys)
PONT WALBY (Neath Port Talbot)
Pont-ar-gothi (Carmarthenshire)
Pont-ar-Hydfer (Powys)
Pont-ar-Ilechau (Carmarthenshire)
Pont-Ebbw (Newport)
Pont-faen (Powys)
Pont-gareg (Pembrokeshire)
PONT-NEDD-FECHAN (Neath Port Talbot)
PONT-RHYD-Y-FEN (Neath Port Talbot)
PONT-Y-BLEW (Wrexham)
Pont-yr-hafod (Pembrokeshire)
PONT-YR-RHYL (Bridgend)
PONTAMMAN (Carmarthenshire)
Pontantwn (Carmarthenshire)
PONTARDAWE (Neath Port Talbot)
PONTARDDULAIS (Swansea)
Pontarsais (Carmarthenshire)
PONTBLYDDYN (Flintshire)
Pontdolgoch (Powys)
PONTEFRACT (West Yorkshire)
PONTELAND (Northumberland)
PONTESBURY (Shropshire)
PONTESBURY HILL (Shropshire)
PONTESFORD (Shropshire)
Pontfadog (Wrexham)
Pontfaen (Pembrokeshire)
PONTHENRY (Carmarthenshire)
Ponthir (Torfaen)
PONTLLANFRAITH (Caerphilly)
PONTLLIW (Swansea)
PONTLOTTYN (Caerphilly)
Pontnewydd (Torfaen)
PONTNEWYNYDD (Torfaen)
PONTOP (Durham)
Pontrhydyrun (Torfaen)
Pontrilas (Hereford & Worcester)
PONTSHILL (Hereford & Worcester)
Pontsticill (Powys)
Pontwelly (Carmarthenshire)
PONTYATES (Carmarthenshire)
PONTYBEREM (Carmarthenshire)
PONTYBODKIN (Flintshire)
PONTYCLUN (Rhondda Cynon Taff)
PONTYCYMER (Bridgend)
Pontyglasier (Pembrokeshire)
PONTYGWAITH (Rhondda Cynon Taff)
Pontygynon (Pembrokeshire)
PONTYMOEL (Torfaen)
Pontypool (Torfaen)
PONTYPOOL ROAD (Torfaen)
PONTYPRIDD (Rhondda Cynon Taff)
PONTYWAUN (Caerphilly)
Pool (West Yorkshire)
Pool Head (Hereford & Worcester)
Pool Quay (Powys)
Pooley Bridge (Cumbria)
POOLFOLD (Staffordshire)
Poolhill (Gloucestershire)
Pooting's (Kent)
PORCHBROOK (Hereford & Worcester)
Porlock (Somerset)
Porlock Weir (Somerset)
Port Carlisle (Cumbria)
Port Clarence (Durham)
Port Einon (Swansea)
PORT MULGRAVE (North Yorkshire)
Port Sunlight (Merseyside)
PORT TALBOT (Neath Port Talbot)
PORT TENNANT (Swansea)
Portbury (Somerset)
Portfield Gate (Pembrokeshire)
PORTH (Rhondda Cynon Taff)
Porth-y-Waen (Shropshire)
Porthcawl (Bridgend)
Porthgain (Pembrokeshire)
PORTHILL (Staffordshire)
Porthkerry (Vale of Glamorgan)
Porthllechog (Anglesey)
Porthwgan (Wrexham)
Porthyrhyd (Carmarthenshire)
Portington (East Riding of Yorkshire)
Portinscale (Cumbria)
Portishead (Somerset)
PORTOBELLO (Tyne & Wear)
PORTOBELLO (West Midlands)
Portskewett (Monmouthshire)

PORTSMOUTH (West Yorkshire)
Portway (Hereford & Worcester)
PORTWAY (West Midlands)
POSENHALL (Shropshire)
Postling (Kent)
POTT SHRIGLEY (Cheshire)
Potten Street (Kent)
Potter Brompton (North Yorkshire)
Potter Somersal (Derbyshire)
Potter's Cross (Staffordshire)
Potter's Forstal (Kent)
Potterhanworth (Lincolnshire)
Potterhanworth Booths (Lincolnshire)
Potters Brook (Lancashire)
POTTERS GREEN (West Midlands)
Potters Marston (Leicestershire)
POTTERTON (West Midlands)
Potto (North Yorkshire)
Poulton (Gloucestershire)
Poulton (Merseyside)
Poulton Priory (Gloucestershire)
Poulton-le-Fylde (Lancashire)
POUND BANK (Hereford & Worcester)
POUND GREEN (Hereford & Worcester)
POUNDFFALD (Swansea)
Poundsbridge (Kent)
Pow Green (Hereford & Worcester)
Powburn (Northumberland)
Powhill (Cumbria)
Powick (Hereford & Worcester)
POYNTON (Cheshire)
Poynton (Shropshire)
Poynton Green (Shropshire)
Poyston Cross (Pembrokeshire)
Prees (Shropshire)
Prees Green (Shropshire)
Prees Heath (Shropshire)
Prees Higher Heath (Shropshire)
Prees Lower Heath (Shropshire)
Preesall (Lancashire)
PREESGWEENE (Shropshire)
Prendwick (Northumberland)
Prenton (Merseyside)
PRESCOT (Merseyside)
Prescott (Shropshire)
Pressen (Northumberland)
Prestatyn (Denbighshire)
Prestbury (Cheshire)
Prestbury (Gloucestershire)
Presteigne (Powys)
Prestleigh (Somerset)
PRESTOLEE (Greater Manchester)
Preston (East Riding of Yorkshire)
Preston (Gloucestershire)
PRESTON (Kent)
Preston (Lancashire)
Preston (Northumberland)
PRESTON (Shropshire)
Preston (Somerset)
Preston Bagot (Warwickshire)
Preston Bowyer (Somerset)
Preston Brockhurst (Shropshire)
Preston Brook (Cheshire)
Preston Green (Warwickshire)
Preston Gubbals (Shropshire)
Preston Montford (Shropshire)
Preston on Stour (Warwickshire)
Preston on the Hill (Cheshire)
Preston on Wye (Hereford & Worcester)
Preston Patrick (Cumbria)
Preston Plucknett (Somerset)
PRESTON STREET (Kent)
Preston upon the Weald Moors (Shropshire)
Preston Wynne (Hereford & Worcester)
PRESTON-UNDER-SCAR (North Yorkshire)
PRESTWICH (Greater Manchester)
PRESTWICK (Northumberland)
Prestwood (Staffordshire)
PRICE TOWN (Bridgend)
Priddy (Somerset)
Priest Hutton (Lancashire)
Priestcliffe (Derbyshire)
Priestcliffe Ditch (Derbyshire)
PRIESTLEY GREEN (West Yorkshire)
Priestweston (Shropshire)
Priestwood Green (Kent)
Primethorpe (Leicestershire)
PRIMROSE HILL (Derbyshire)
Primrose Hill (Lancashire)
PRIMROSE HILL (West Midlands)
Princes Gate (Pembrokeshire)

Princethorpe (Warwickshire)
Prion (Denbighshire)
PRIOR RIGG (Cumbria)
Priors Halton (Shropshire)
Priors Hardwick (Warwickshire)
Priors Marston (Warwickshire)
Priors Norton (Gloucestershire)
Priory Wood (Hereford & Worcester)
Prisk (Vale of Glamorgan)
Priston (Somerset)
PROSPECT (Cumbria)
PROVIDENCE (Somerset)
PRUDHOE (Northumberland)
PUBLOW (Somerset)
Puckington (Somerset)
PUCKLECHURCH (Gloucestershire)
Puckrup (Gloucestershire)
Puddinglake (Cheshire)
Puddington (Cheshire)
PUDSEY (West Yorkshire)
Puleston (Shropshire)
Pulford (Cheshire)
Pullens Green (Gloucestershire)
PULLEY (Shropshire)
Pumsaint (Carmarthenshire)
Puncheston (Pembrokeshire)
Puriton (Somerset)
Purlogue (Shropshire)
Purshall Green (Hereford & Worcester)
Purslow (Shropshire)
PURSTON JAGLIN (West Yorkshire)
Purtington (Somerset)
Purton (Gloucestershire)
Putley (Hereford & Worcester)
Putley Green (Hereford & Worcester)
Putloe (Gloucestershire)
Puxton (Somerset)
PWLL (Carmarthenshire)
Pwll Trap (Carmarthenshire)
PWLL-DU (Monmouthshire)
Pwll-glas (Denbighshire)
PWLL-Y-GLAW (Neath Port Talbot)
Pwllcrochan (Pembrokeshire)
Pwllgloyw (Powys)
Pwllmeyric (Monmouthshire)
PYE BRIDGE (Derbyshire)
Pye Corner (Newport)
PYE GREEN (Staffordshire)
PYLE (Bridgend)
Pyleigh (Somerset)
Pylle (Somerset)

Q

Quabbs (Shropshire)
Quadring (Lincolnshire)
Quadring Eaudike (Lincolnshire)
QUAKER'S YARD (Merthyr Tydfil)
QUAKING HOUSES (Durham)
Quarndon (Derbyshire)
Quarrington (Lincolnshire)
QUARRINGTON HILL (Durham)
QUARRY BANK (Cheshire)
QUARRY BANK (West Midlands)
Quatford (Shropshire)
QUATT (Shropshire)
QUEBEC (Durham)
Quedgeley (Gloucestershire)
Queen Camel (Somerset)
Queen Charlton (Somerset)
Queen Street (Kent)
Queen's Head (Shropshire)
Queenborough (Kent)
Queenhill (Hereford & Worcester)
QUEENSBURY (West Yorkshire)
QUEENSFERRY (Flintshire)
Queniborough (Leicestershire)
Quenington (Gloucestershire)
QUERNMORE (Lancashire)
QUERNMORE PARK HALL (Lancashire)
QUESLETT (West Midlands)
Quina Brook (Shropshire)
QUINTON (West Midlands)
Quixhall (Staffordshire)
Quorndon (Leicestershire)

R

Rabbit's Cross (Kent)
Raby (Cumbria)
RABY (Merseyside)

See paras. 2 and 4 of the User Guide 2003 if you can't find your place name.

Radbourne (Derbyshire)
RADCLIFFE (Greater Manchester)
RADCLIFFE (Northumberland)
RADCLIFFE ON TRENT (Nottinghamshire)
Raddington (Somerset)
Radford Semele (Warwickshire)
Radlet (Somerset)
Radmore Green (Cheshire)
RADSTOCK (Somerset)
Radway (Warwickshire)
Radway Green (Cheshire)
Radyr (Cardiff)
RAF College (Cranwell) (Lincolnshire)
RAGDALE (Leicestershire)
Ragdon (Shropshire)
Raglan (Monmouthshire)
RAGNALL (Nottinghamshire)
Rainbow Hill (Hereford & Worcester)
RAINFORD (Merseyside)
Rainham (Kent)
RAINHILL (Merseyside)
RAINHILL STOOPS (Merseyside)
RAINOW (Cheshire)
RAINSOUGH (Greater Manchester)
Rainton (North Yorkshire)
RAINWORTH (Nottinghamshire)
Raisbeck (Cumbria)
RAISE (Cumbria)
Raisthorpe (North Yorkshire)
Raithby (Lincolnshire)
RAITHWAITE (North Yorkshire)
RAKEWOOD (Greater Manchester)
Ram (Carmarthenshire)
RAM HILL (Gloucestershire)
Ram Lane (Kent)
Rampside (Cumbria)
Rampton (Nottinghamshire)
RAMSBOTTOM (Greater Manchester)
Ramsden (Hereford & Worcester)
Ramsgate (Kent)
RAMSGILL (North Yorkshire)
Ramshaw (Durham)
Ramshope (Northumberland)
Ramshorn (Staffordshire)
Ranby (Lincolnshire)
RANBY (Nottinghamshire)
Rand (Lincolnshire)
Randwick (Gloucestershire)
Rangemore (Staffordshire)
RANGEWORTHY (Gloucestershire)
RANN (Lancashire)
Ranscombe (Somerset)
RANSKILL (Nottinghamshire)
Ranton (Staffordshire)
Ranton Green (Staffordshire)
Rapps (Somerset)
Rashwood (Hereford & Worcester)
Raskelf (North Yorkshire)
RASSAU (Blaenau Gwent)
RASTRICK (West Yorkshire)
Ratby (Leicestershire)
Ratcliffe Culey (Leicestershire)
Ratcliffe on Soar (Nottinghamshire)
Ratcliffe on the Wreake (Leicestershire)
Rathmell (North Yorkshire)
Ratley (Warwickshire)
RATLING (Kent)
Ratlinghope (Shropshire)
RATTEN ROW (Cumbria)
Ratten Row (Lancashire)
RAUGHTON (Cumbria)
RAUGHTON HEAD (Cumbria)
Raven Meols (Merseyside)
RAVENFIELD (South Yorkshire)
Ravenglass (Cumbria)
Ravenhills Green (Hereford & Worcester)
Ravenscar (North Yorkshire)
RAVENSCLIFFE (Staffordshire)
RAVENSHEAD (Nottinghamshire)
Ravensmoor (Cheshire)
RAVENSTHORPE (West Yorkshire)
RAVENSTONE (Leicestershire)
RAVENSTONEDALE (Cumbria)
Ravensworth (North Yorkshire)
RAW (North Yorkshire)
RAWCLIFFE (East Riding of Yorkshire)
Rawcliffe (North Yorkshire)
RAWCLIFFE BRIDGE (East Riding of Yorkshire)
RAWDON (West Yorkshire)
Rawling Street (Kent)
RAWMARSH (South Yorkshire)

RAWNSLEY (Staffordshire)
RAWTENSTALL (Lancashire)
RAYLEES (Northumberland)
Rea (Gloucestershire)
READ (Lancashire)
Reading Street (Kent)
REAGILL (Cumbria)
Rearsby (Leicestershire)
Rease Heath (Shropshire)
Reculver (Kent)
RED BULL (Cheshire)
RED DIAL (Cumbria)
Red Hill (Warwickshire)
RED LUMB (Greater Manchester)
RED ROCK (Greater Manchester)
Red Roses (Carmarthenshire)
RED ROW (Tyne & Wear)
RED STREET (Staffordshire)
Red Wharf Bay (Anglesey)
REDBERTH (Pembrokeshire)
Redbourne (Lincolnshire)
Redbrook (Gloucestershire)
Redbrook (Wrexham)
Redbrook Street (Kent)
REDBURN (Northumberland)
Redcar (North Yorkshire)
REDDISH (Greater Manchester)
Redditch (Hereford & Worcester)
REDESMOUTH (Northumberland)
REDFIELD (Bristol)
REDGATE (Rhondda Cynon Taff)
Redhill (Somerset)
Redland (Bristol)
Redlynch (Somerset)
Redmain (Cumbria)
REDMARLEY (Hereford & Worcester)
Redmarley D'Abitot (Gloucestershire)
Redmarshall (Durham)
REDMILE (Leicestershire)
REDMIRE (North Yorkshire)
Rednal (Shropshire)
Rednal (West Midlands)
Redstone Cross (Pembrokeshire)
REDVALES (Greater Manchester)
Redwick (Gloucestershire)
Redwick (Newport)
Redworth (Durham)
Reedness (East Riding of Yorkshire)
REEDS HOLME (Lancashire)
Reepham (Lincolnshire)
Reeth (North Yorkshire)
REEVES GREEN (West Midlands)
Reighton (North Yorkshire)
Rempstone (Nottinghamshire)
Rendcomb (Gloucestershire)
RENISHAW (Derbyshire)
RENNINGTON (Northumberland)
RENWICK (Cumbria)
Repton (Derbyshire)
RESOLVEN (Neath Port Talbot)
RETFORD (Nottinghamshire)
Revesby (Lincolnshire)
REYNALTON (Pembrokeshire)
Reynoldston (Swansea)
Rhadyr (Monmouthshire)
Rhandirmwyn (Carmarthenshire)
Rhayader (Powys)
Rhes-y-cae (Flintshire)
Rhewl (Denbighshire)
RHEWL MOSTYN (Flintshire)
RHEWL-FAWR (Flintshire)
RHIGOS (Rhondda Cynon Taff)
Rhiwbina (Cardiff)
Rhiwderyn (Newport)
RHIWINDER (Rhondda Cynon Taff)
Rhiwlas (Powys)
RHIWSAESON (Cardiff)
Rhode (Somerset)
Rhoden Green (Kent)
RHODES (Greater Manchester)
Rhodes Minnis (Kent)
RHODESIA (Nottinghamshire)
Rhodiad-y-brenin (Pembrokeshire)
Rhoose (Vale of Glamorgan)
Rhos (Carmarthenshire)
Rhos (Denbighshire)
RHOS (Neath Port Talbot)
Rhos (Powys)
Rhos Lligwy (Anglesey)
Rhos y-brithdir (Powys)
Rhos-y-meirch (Powys)

Rhosbeirio (Anglesey)
Rhoscefnhir (Anglesey)
Rhoscolyn (Anglesey)
Rhoscrowther (Pembrokeshire)
Rhosesmor (Flintshire)
Rhosgoch (Anglesey)
Rhosgoch (Powys)
Rhoshill (Pembrokeshire)
RHOSLLANERCHRUGOG (Wrexham)
Rhosmaen (Carmarthenshire)
Rhosmeirch (Anglesey)
Rhosneigr (Anglesey)
RHOSNESNI (Wrexham)
RHOSROBIN (Wrexham)
Rhossili (Swansea)
RHOSTYLLEN (Wrexham)
Rhosybol (Anglesey)
RHOSYGADFA (Shropshire)
RHOSYMEDRE (Wrexham)
Rhuallt (Denbighshire)
Rhuddall Heath (Cheshire)
Rhuddlan (Denbighshire)
Rhulen (Powys)
Rhyd-y-meirch (Monmouthshire)
Rhydargaeau (Carmarthenshire)
Rhydcymerau (Carmarthenshire)
Rhydd (Hereford & Worcester)
RHYDDING (Neath Port Talbot)
Rhydspence (Hereford & Worcester)
Rhydtalog (Flintshire)
Rhydwyn (Anglesey)
Rhydycroesau (Shropshire)
RHYDYFELIN (Rhondda Cynon Taff)
RHYDYFRO (Neath Port Talbot)
Rhydymwyn (Flintshire)
Rhyl (Denbighshire)
RHYMNEY (Caerphilly)
RIBBESFORD (Hereford & Worcester)
Ribbleton (Lancashire)
Ribby (Lancashire)
Ribchester (Lancashire)
Riber (Derbyshire)
Riby (Lincolnshire)
RICCALL (North Yorkshire)
Rich's Holford (Somerset)
Richards Castle (Hereford & Worcester)
Richmond (North Yorkshire)
RICHMOND (South Yorkshire)
RICKERSCOTE (Staffordshire)
Rickford (Somerset)
RIDDINGS (Cumbria)
RIDDINGS (Derbyshire)
RIDDLESDEN (West Yorkshire)
Ridge (Somerset)
RIDGE LANE (Warwickshire)
Ridge Row (Kent)
Ridgebourne (Powys)
Ridgehill (Somerset)
RIDGEWAY (Derbyshire)
Ridgeway (Hereford & Worcester)
Ridgeway Cross (Hereford & Worcester)
RIDING MILL (Northumberland)
Ridley (Kent)
RIDLEY (Northumberland)
Ridley Green (Cheshire)
RIDSDALE (Northumberland)
RIEVAULX (North Yorkshire)
RIGMADON PARK (Cumbria)
Rigsby (Lincolnshire)
RILEY GREEN (Lancashire)
Rileyhill (Staffordshire)
Rillington (North Yorkshire)
Rimington (Lancashire)
Rimpton (Somerset)
Rimswell (East Riding of Yorkshire)
Rinaston (Pembrokeshire)
Rindleford (Shropshire)
RING O'BELLS (Lancashire)
RINGINGLOW (Derbyshire)
Ringlestone (Kent)
RINGLEY (Greater Manchester)
Ringwould (Kent)
RIPLEY (Derbyshire)
Ripley (North Yorkshire)
Riplingham (East Riding of Yorkshire)
Ripon (North Yorkshire)
Rippingale (Lincolnshire)
Ripple (Hereford & Worcester)
RIPPLE (Kent)
Ripponden (West Yorkshire)
Risbury (Hereford & Worcester)

See paras. 2 and 4 of the User Guide 2003 if you can't find your place name.

Risby (Lincolnshire)
RISCA (Caerphilly)
Rise (East Riding of Yorkshire)
Risedown (Kent)
Risegate (Lincolnshire)
Riseholme (Lincolnshire)
RISEHOW (Cumbria)
RISHTON (Lancashire)
Rishworth (West Yorkshire)
RISING BRIDGE (Lancashire)
RISLEY (Cheshire)
Risley (Derbyshire)
RISPLITH (North Yorkshire)
River (Kent)
Riverhead (Kent)
RIVINGTON (Lancashire)
ROADHEAD (Cumbria)
Roadwater (Somerset)
Roath (Cardiff)
ROBERTTOWN (West Yorkshire)
Robeston Wathen (Pembrokeshire)
ROBIN HILL (Staffordshire)
ROBIN HOOD (Lancashire)
ROBIN HOOD (West Yorkshire)
Robin Hood's Bay (North Yorkshire)
Roby (Merseyside)
ROBY MILL (Lancashire)
Rocester (Staffordshire)
ROCH (Pembrokeshire)
ROCH GATE (Pembrokeshire)
ROCHDALE (Greater Manchester)
Rochester (Kent)
ROCHESTER (Northumberland)
Rochford (Hereford & Worcester)
ROCK (Hereford & Worcester)
ROCK (Neath Port Talbot)
ROCK (Northumberland)
Rock Ferry (Merseyside)
Rock Hill (Hereford & Worcester)
Rockcliffe (Cumbria)
Rockcliffe Cross (Cumbria)
Rockfield (Monmouthshire)
Rockgreen (Shropshire)
Rockhampton (Gloucestershire)
Rockhill (Shropshire)
ROCKLEY (Nottinghamshire)
ROCKLIFFE (Lancashire)
Rockwell Green (Somerset)
Rodborough (Gloucestershire)
Rodd (Hereford & Worcester)
Roddam (Northumberland)
RODDYMOOR (Durham)
Rode (Somerset)
Rode Heath (Cheshire)
Roden (Shropshire)
Rodhuish (Somerset)
Rodington (Shropshire)
Rodington Heath (Shropshire)
Rodley (Gloucestershire)
RODLEY (West Yorkshire)
Rodmarton (Gloucestershire)
Rodmersham (Kent)
Rodmersham Green (Kent)
Rodney Stoke (Somerset)
Rodsley (Derbyshire)
Rodway (Somerset)
ROE CROSS (Greater Manchester)
ROE GREEN (Greater Manchester)
Roecliffe (North Yorkshire)
Roger Ground (Cumbria)
Rogerstone (Newport)
Rogiet (Monmouthshire)
ROKER (Tyne & Wear)
Rolleston (Leicestershire)
Rolleston (Nottinghamshire)
Rolleston (Staffordshire)
Rolston (East Riding of Yorkshire)
Rolstone (Somerset)
Rolvenden (Kent)
Rolvenden Layne (Kent)
Romaldkirk (Durham)
Romanby (North Yorkshire)
Romden Castle (Kent)
ROMILEY (Greater Manchester)
Romney Street (Kent)
Romsley (Hereford & Worcester)
ROMSLEY (Shropshire)
Rookhope (Durham)
Rooks Bridge (Somerset)
Rooks Nest (Somerset)
Rookwith (North Yorkshire)

Roos (East Riding of Yorkshire)
Roose (Cumbria)
Roosebeck (Cumbria)
Ropsley (Lincolnshire)
Rorrington (Shropshire)
ROSE HILL (Lancashire)
Roseacre (Lancashire)
Rosebush (Pembrokeshire)
ROSEDALE ABBEY (North Yorkshire)
Roseden (Northumberland)
Rosehill (Shropshire)
Rosemarket (Pembrokeshire)
Roseworth (Durham)
Rosgill (Cumbria)
ROSLEY (Cumbria)
ROSLISTON (Derbyshire)
Ross (Northumberland)
Ross-on-Wye (Hereford & Worcester)
ROSSETT (Wrexham)
Rossett Green (North Yorkshire)
ROSSINGTON (South Yorkshire)
Rostherne (Cheshire)
Rosthwaite (Cumbria)
Roston (Derbyshire)
ROTHBURY (Northumberland)
ROTHERBY (Leicestershire)
ROTHERHAM (South Yorkshire)
Rothley (Leicestershire)
ROTHLEY (Northumberland)
Rothwell (Lincolnshire)
ROTHWELL (West Yorkshire)
ROTHWELL HAIGH (West Yorkshire)
Rotsea (East Riding of Yorkshire)
ROTTINGTON (Cumbria)
ROUGH CLOSE (Staffordshire)
Rough Common (Kent)
ROUGHLEE (Lancashire)
Roughley (West Midlands)
Roughton (Lincolnshire)
Roughton (Shropshire)
Roughway (Kent)
Round Street (Kent)
Roundham (Somerset)
ROUNDHAY (West Yorkshire)
ROUNDS GREEN (West Midlands)
Rous Lench (Hereford & Worcester)
Routenbeck (Cumbria)
Routh (East Riding of Yorkshire)
ROW (Cumbria)
ROWARTH (Derbyshire)
Rowberrow (Somerset)
Rowfield (Derbyshire)
ROWFOOT (Northumberland)
Rowford (Somerset)
Rowington (Warwickshire)
Rowland (Derbyshire)
ROWLAND'S GILL (Tyne & Wear)
ROWLEY (Durham)
Rowley (East Riding of Yorkshire)
Rowley (Shropshire)
ROWLEY GREEN (West Midlands)
ROWLEY HILL (West Yorkshire)
ROWLEY REGIS (West Midlands)
Rowlstone (Hereford & Worcester)
Rowney Green (Hereford & Worcester)
ROWRAH (Cumbria)
Rows of Trees (Cheshire)
Rowsley (Derbyshire)
Rowston (Lincolnshire)
ROWTHORNE (Derbyshire)
Rowton (Cheshire)
ROWTON (Shropshire)
Roxby (Lincolnshire)
ROXBY (North Yorkshire)
Royal Oak (Durham)
Royal Oak (Lancashire)
Royal's Green (Cheshire)
ROYDHOUSE (West Yorkshire)
ROYSTON (South Yorkshire)
ROYTON (Greater Manchester)
RUABON (Wrexham)
RUARDEAN (Gloucestershire)
RUARDEAN HILL (Gloucestershire)
RUARDEAN WOODSIDE (Gloucestershire)
Rubery (Hereford & Worcester)
Ruckcroft (Cumbria)
Ruckhall Common (Hereford & Worcester)
Ruckinge (Kent)
Ruckland (Lincolnshire)
RUCKLEY (Shropshire)
Rudby (North Yorkshire)

RUDCHESTER (Northumberland)
RUDDINGTON (Nottinghamshire)
Ruddle (Gloucestershire)
Rudford (Gloucestershire)
Rudge (Somerset)
Rudgeway (Gloucestershire)
Rudhall (Hereford & Worcester)
Rudheath (Cheshire)
RUDRY (Caerphilly)
Rudston (East Riding of Yorkshire)
Rudyard (Staffordshire)
Rufford (Lancashire)
Rufforth (North Yorkshire)
Rug (Denbighshire)
Rugby (Warwickshire)
RUGELEY (Staffordshire)
Ruishton (Somerset)
RUMBY HILL (Durham)
Rumney (Cardiff)
Rumwell (Somerset)
Runcorn (Cheshire)
RUNNING WATERS (Durham)
Runnington (Somerset)
RUNSHAW MOOR (Lancashire)
Runswick (North Yorkshire)
Rush Green (Cheshire)
Rushall (Hereford & Worcester)
RUSHALL (West Midlands)
Rushbury (Shropshire)
Rushenden (Kent)
Rushock (Hereford & Worcester)
RUSHOLME (Greater Manchester)
Rushton (Cheshire)
RUSHTON (Shropshire)
Rushton Spencer (Staffordshire)
Rushwick (Hereford & Worcester)
RUSHYFORD (Durham)
Ruskington (Lincolnshire)
Rusland (Cumbria)
RUSPIDGE (Gloucestershire)
Rusthall (Kent)
Ruston (North Yorkshire)
Ruston Parva (East Riding of Yorkshire)
Ruswarp (North Yorkshire)
Ruthall (Shropshire)
Ruthin (Denbighshire)
Ruthwaite (Cumbria)
Ruxton Green (Hereford & Worcester)
Ruyton-XI-Towns (Shropshire)
RYAL (Northumberland)
Ryall (Hereford & Worcester)
Ryarsh (Kent)
Rydal (Cumbria)
Rye Cross (Hereford & Worcester)
Rye Street (Hereford & Worcester)
Ryebank (Shropshire)
Ryeford (Hereford & Worcester)
RYHILL (West Yorkshire)
RYHOPE (Tyne & Wear)
RYLAH (Derbyshire)
Ryland (Lincolnshire)
RYLANDS (Nottinghamshire)
RYLSTONE (North Yorkshire)
RYTHER (North Yorkshire)
Ryton (North Yorkshire)
RYTON (Shropshire)
RYTON (Tyne & Wear)
Ryton (Warwickshire)
RYTON WOODSIDE (Tyne & Wear)
RYTON-ON-DUNSMORE (Warwickshire)

S

SABDEN (Lancashire)
SACRISTON (Durham)
Sadberge (Durham)
Saddington (Leicestershire)
Sadgill (Cumbria)
Sageston (Pembrokeshire)
Saighton (Cheshire)
Saintbury (Gloucestershire)
Sale (Greater Manchester)
Sale Green (Hereford & Worcester)
Saleby (Lincolnshire)
Salem (Carmarthenshire)
Salesbury (Lancashire)
SALFORD (Greater Manchester)
Salford Priors (Warwickshire)
Salkeld Dykes (Cumbria)
Salmonby (Lincolnshire)
Salperton (Gloucestershire)

Salt (Staffordshire)
Salt Cotes (Cumbria)
SALTA (Cumbria)
SALTAIRE (West Yorkshire)
Saltburn-by-the-Sea (North Yorkshire)
Saltby (Leicestershire)
Saltcoats (Cumbria)
Saltcotes (Lancashire)
SALTERBECK (Cumbria)
Salterforth (Lancashire)
Salterswall (Cheshire)
Saltfleet (Lincolnshire)
Saltfleetby All Saints (Lincolnshire)
Saltfleetby St. Clements (Lincolnshire)
Saltfleetby St. Peter (Lincolnshire)
Saltford (Somerset)
Saltley (West Midlands)
Saltmarsh (Newport)
Saltmarshe (East Riding of Yorkshire)
SALTNEY (Flintshire)
Salton (North Yorkshire)
SALTWICK (Northumberland)
Saltwood (Kent)
Salwarpe (Hereford & Worcester)
Sambourne (Warwickshire)
Sambrook (Shropshire)
Samlesbury (Lancashire)
Samlesbury Bottoms (Lancashire)
Sampford Arundel (Somerset)
Sampford Brett (Somerset)
Sampford Moor (Somerset)
Sancton (East Riding of Yorkshire)
Sand (Somerset)
SAND HILLS (West Yorkshire)
Sand Hole (East Riding of Yorkshire)
Sand Hutton (North Yorkshire)
Sand Side (Cumbria)
SANDAL MAGNA (West Yorkshire)
SANDALE (Cumbria)
Sandbach (Cheshire)
Sandford (Cumbria)
Sandford (Shropshire)
Sandford (Somerset)
Sandgate (Kent)
SANDHILL (South Yorkshire)
SANDHILLS (West Midlands)
SANDHOE (Northumberland)
Sandholme (East Riding of Yorkshire)
Sandholme (Lincolnshire)
Sandhurst (Gloucestershire)
Sandhurst (Kent)
Sandhurst Cross (Kent)
Sandhutton (North Yorkshire)
Sandiacre (Derbyshire)
Sandilands (Lincolnshire)
Sandiway (Cheshire)
Sandling (Kent)
Sandlow Green (Cheshire)
Sandon (Staffordshire)
Sandon Bank (Staffordshire)
Sandsend (North Yorkshire)
Sandside (Cumbria)
Sandtoft (Lincolnshire)
Sandway (Kent)
SANDWICH (Kent)
Sandwick (Cumbria)
SANDWITH (Cumbria)
SANDWITH NEWTOWN (Cumbria)
Sandy Bank (Lincolnshire)
Sandy Cross (Hereford & Worcester)
Sandy Haven (Pembrokeshire)
SANDY LANE (West Yorkshire)
SANDY LANE (Wrexham)
SANDYCROFT (Flintshire)
Sandylands (Lancashire)
Sandylane (Staffordshire)
Sandylane (Swansea)
Sandyway (Hereford & Worcester)
SANKEY BRIDGES (Cheshire)
Sankyn's Green (Hereford & Worcester)
Santon (Cumbria)
Santon Bridge (Cumbria)
Sapcote (Leicestershire)
Sapey Common (Hereford & Worcester)
Sapperton (Derbyshire)
Sapperton (Gloucestershire)
Sapperton (Lincolnshire)
Saracen's Head (Lincolnshire)
SARN (Bridgend)
Sarn (Powys)
Sarn-wen (Powys)

Sarnau (Carmarthenshire)
Sarnau (Powys)
Sarnesfield (Hereford & Worcester)
SARON (Carmarthenshire)
Sarre (Kent)
SATLEY (Durham)
Satmar (Kent)
SATRON (North Yorkshire)
Satterthwaite (Cumbria)
Saul (Gloucestershire)
SAUNDBY (Nottinghamshire)
SAUNDERSFOOT (Pembrokeshire)
Sausthorpe (Lincolnshire)
Saverley Green (Staffordshire)
SAVILE TOWN (West Yorkshire)
Sawbridge (Warwickshire)
Sawdon (North Yorkshire)
Sawley (Derbyshire)
Sawley (Lancashire)
SAWLEY (North Yorkshire)
Saxby (Leicestershire)
Saxby (Lincolnshire)
Saxby All Saints (Lincolnshire)
SAXELBYE (Leicestershire)
Saxilby (Lincolnshire)
SAXONDALE (Nottinghamshire)
Saxton (North Yorkshire)
SCACKLETON (North Yorkshire)
SCAFTWORTH (Nottinghamshire)
Scagglethorpe (North Yorkshire)
Scalby (East Riding of Yorkshire)
Scalby (North Yorkshire)
SCALE HOUSES (Cumbria)
Scaleby (Cumbria)
Scalebyhill (Cumbria)
Scales (Cumbria)
Scales (Lancashire)
Scalesceugh (Cumbria)
SCALFORD (Leicestershire)
Scaling (North Yorkshire)
Scaling Dam (North Yorkshire)
Scamblesby (Lincolnshire)
Scammonden (West Yorkshire)
Scampston (North Yorkshire)
Scampton (Lincolnshire)
Scapegoat Hill (West Yorkshire)
Scarborough (North Yorkshire)
SCARCLIFFE (Derbyshire)
Scarcroft (West Yorkshire)
Scarcroft Hill (West Yorkshire)
Scargill (Durham)
Scarisbrick (Lancashire)
Scarness (Cumbria)
Scarrington (Nottinghamshire)
Scarth Hill (Lancashire)
Scarthingwell (North Yorkshire)
Scartho (Lincolnshire)
Scawby (Lincolnshire)
SCAWSBY (South Yorkshire)
SCAWTHORPE (South Yorkshire)
SCAWTON (North Yorkshire)
Scethrog (Powys)
SCHOLAR GREEN (Staffordshire)
SCHOLES (Greater Manchester)
SCHOLES (South Yorkshire)
SCHOLES (West Yorkshire)
SCHOLEY HILL (West Yorkshire)
School Aycliffe (Durham)
School Green (Cheshire)
SCHOOL GREEN (West Yorkshire)
SCISSETT (West Yorkshire)
Scleddau (Pembrokeshire)
SCOFTON (Nottinghamshire)
Scopwick (Lincolnshire)
Scorborough (East Riding of Yorkshire)
Scorton (Lancashire)
Scorton (North Yorkshire)
SCOT HAY (Staffordshire)
SCOT LANE END (Greater Manchester)
SCOT'S GAP (Northumberland)
Scotby (Cumbria)
Scotch Corner (North Yorkshire)
Scotforth (Lancashire)
Scothern (Lincolnshire)
Scotland (Lincolnshire)
SCOTLAND (West Yorkshire)
SCOTLAND GATE (Tyne & Wear)
Scotsdike (Cumbria)
SCOTSWOOD (Tyne & Wear)
Scotter (Lincolnshire)
Scotterthorpe (Lincolnshire)

Scottlethorpe (Lincolnshire)
Scotton (Lincolnshire)
SCOTTON (North Yorkshire)
Scounslow Green (Staffordshire)
SCOUTHEAD (Greater Manchester)
Scrafield (Lincolnshire)
Scrainwood (Northumberland)
Scrane End (Lincolnshire)
Scraptoft (Leicestershire)
Scrayingham (North Yorkshire)
Scredington (Lincolnshire)
Scremby (Lincolnshire)
SCREMERSTON (Northumberland)
Screveton (Nottinghamshire)
Scrivelby (Lincolnshire)
Scriven (North Yorkshire)
SCROOBY (Nottinghamshire)
Scropton (Derbyshire)
Scrub Hill (Lincolnshire)
SCRUTON (North Yorkshire)
SCUGGATE (Cumbria)
Scunthorpe (Lincolnshire)
Scurlage (Swansea)
Sea (Somerset)
Sea Mills (Bristol)
SEABRIDGE (Staffordshire)
Seabrook (Kent)
SEABURN (Tyne & Wear)
Seacombe (Merseyside)
Seacroft (Lincolnshire)
SEACROFT (West Yorkshire)
Seadyke (Lincolnshire)
Seaforth (Merseyside)
SEAGRAVE (Leicestershire)
SEAHAM (Durham)
SEAHOUSES (Northumberland)
Seal (Kent)
Seamer (North Yorkshire)
Searby (Lincolnshire)
Seasalter (Kent)
Seascale (Cumbria)
Seathwaite (Cumbria)
Seatle (Cumbria)
Seatoller (Cumbria)
SEATON (Cumbria)
SEATON (Durham)
Seaton (East Riding of Yorkshire)
SEATON (Kent)
SEATON (Northumberland)
SEATON BURN (Tyne & Wear)
Seaton Carew (Durham)
SEATON DELAVAL (Northumberland)
Seaton Ross (East Riding of Yorkshire)
SEATON SLUICE (Northumberland)
SEAVE GREEN (North Yorkshire)
Seaville (Cumbria)
Seavington St. Mary (Somerset)
Seavington St. Michael (Somerset)
Sebastopol (Torfaen)
SEBERGHAM (Cumbria)
SECKINGTON (Warwickshire)
Sedbergh (Cumbria)
Sedbury (Gloucestershire)
SEDBUSK (North Yorkshire)
Sedgeberrow (Hereford & Worcester)
Sedgebrook (Lincolnshire)
SEDGEFIELD (Durham)
SEDGLEY (West Midlands)
SEDGLEY PARK (Greater Manchester)
Sedgwick (Cumbria)
Seed (Kent)
Sefton (Merseyside)
Sefton Town (Merseyside)
SEGHILL (Northumberland)
Seighford (Staffordshire)
Seisdon (Staffordshire)
SELATTYN (Shropshire)
SELBY (North Yorkshire)
Sellack (Hereford & Worcester)
Sellick's Green (Somerset)
Sellindge (Kent)
Selling (Kent)
Selly Oak (West Midlands)
Selside (Cumbria)
Selside (North Yorkshire)
Selsted (Kent)
SELSTON (Nottinghamshire)
Selworthy (Somerset)
SENGHENYDD (Caerphilly)
Sennybridge (Powys)
SERLBY (Nottinghamshire)

See paras. 2 and 4 of the User Guide 2003 if you can't find your place name.

SESSAY (North Yorkshire)
Settle (North Yorkshire)
Settlingstones (Northumberland)
Settrington (North Yorkshire)
Seven Ash (Somerset)
SEVEN SISTERS (Neath Port Talbot)
Seven Springs (Gloucestershire)
Seven Wells (Gloucestershire)
Sevenhampton (Gloucestershire)
Sevenoaks (Kent)
Sevenoaks Weald (Kent)
Severn Beach (Gloucestershire)
Severn Stoke (Hereford & Worcester)
Sevington (Kent)
Sewerby (East Riding of Yorkshire)
SEWSTERN (Leicestershire)
Sexhow (North Yorkshire)
Sezincote (Gloucestershire)
Shackerstone (Leicestershire)
Shacklecross (Derbyshire)
Shade (West Yorkshire)
SHADFORTH (Durham)
Shadoxhurst (Kent)
SHADWELL (West Yorkshire)
SHAFTHOLME (South Yorkshire)
SHAFTON (South Yorkshire)
SHAFTON TWO GATES (South Yorkshire)
Shallowford (Staffordshire)
Shalmsford Street (Kent)
Shangton (Leicestershire)
SHANKHOUSE (Northumberland)
Shap (Cumbria)
Shapwick (Somerset)
Shard End (West Midlands)
Shardlow (Derbyshire)
SHARESHILL (Staffordshire)
SHARLSTON (West Yorkshire)
SHARLSTON COMMON (West Yorkshire)
Sharman's Cross (West Midlands)
Sharnal Street (Kent)
SHARNEYFORD (Lancashire)
Sharnford (Leicestershire)
Sharoe Green (Lancashire)
Sharow (North Yorkshire)
Sharperton (Northumberland)
Sharpness (Gloucestershire)
Sharpway Gate (Hereford & Worcester)
SHATTERFORD (Hereford & Worcester)
SHATTERLING (Kent)
Shavington (Cheshire)
SHAW (Greater Manchester)
SHAW (West Yorkshire)
Shaw Common (Gloucestershire)
SHAW GREEN (Lancashire)
Shaw Green (North Yorkshire)
SHAW HILL (Lancashire)
Shaw Mills (North Yorkshire)
Shawbury (Shropshire)
SHAWCLOUGH (Greater Manchester)
Shawdon Hill (Northumberland)
Shawell (Leicestershire)
SHAWFORTH (Lancashire)
Shearsby (Leicestershire)
Shearston (Somerset)
Shebdon (Staffordshire)
Sheen (Derbyshire)
SHEEP HILL (Durham)
SHEEP-RIDGE (West Yorkshire)
SHEEPBRIDGE (Derbyshire)
SHEEPSCAR (West Yorkshire)
Sheepscombe (Gloucestershire)
SHEEPWASH (Northumberland)
Sheepway (Somerset)
Sheepy Magna (Leicestershire)
Sheepy Parva (Leicestershire)
Sheerness (Kent)
SHEFFIELD (South Yorkshire)
Sheinton (Shropshire)
Shelderton (Shropshire)
Sheldon (Derbyshire)
Sheldon (West Midlands)
Sheldwich (Kent)
Sheldwich Lees (Kent)
SHELF (West Yorkshire)
Shelfield (Warwickshire)
SHELFIELD (West Midlands)
Shelfield Green (Warwickshire)
SHELFORD (Nottinghamshire)
Shelford (Warwickshire)
SHELLEY (West Yorkshire)
SHELLEY FAR BANK (West Yorkshire)

Shelsley Beauchamp (Hereford & Worcester)
Shelsley Walsh (Hereford & Worcester)
Shelton (Nottinghamshire)
SHELTON (Shropshire)
Shelton Lock (Derbyshire)
SHELTON UNDER HARLEY (Staffordshire)
Shelve (Shropshire)
Shelwick (Hereford & Worcester)
Shenmore (Hereford & Worcester)
Shenstone (Hereford & Worcester)
Shenstone (Staffordshire)
Shenstone Woodend (Staffordshire)
Shenton (Leicestershire)
Shepeau Stow (Lincolnshire)
Shepherds Patch (Gloucestershire)
SHEPHERDSWELL (Kent)
SHEPLEY (West Yorkshire)
Shepperdine (Gloucestershire)
SHEPSHED (Leicestershire)
Shepton Beauchamp (Somerset)
Shepton Mallet (Somerset)
Shepton Montague (Somerset)
Shepway (Kent)
SHERATON (Durham)
Sherborne (Gloucestershire)
SHERBORNE (Somerset)
Sherbourne (Warwickshire)
SHERBURN (Durham)
Sherburn (North Yorkshire)
SHERBURN HILL (Durham)
SHERBURN IN ELMET (North Yorkshire)
SHERFIN (Lancashire)
Sheriff Hutton (North Yorkshire)
SHERIFFHALES (Shropshire)
SHERWOOD (Nottinghamshire)
SHEVINGTON (Greater Manchester)
SHEVINGTON MOOR (Greater Manchester)
SHEVINGTON VALE (Greater Manchester)
SHIBDEN HEAD (West Yorkshire)
Shidlaw (Northumberland)
SHIFNAL (Shropshire)
SHILBOTTLE (Northumberland)
SHILDON (Durham)
Shillmoor (Northumberland)
Shilton (Warwickshire)
SHINCLIFFE (Durham)
SHINEY ROW (Tyne & Wear)
Shipbourne (Kent)
Shipbrookhill (Cheshire)
Shipham (Somerset)
SHIPLEY (Derbyshire)
Shipley (Shropshire)
SHIPLEY (West Yorkshire)
Shipley Hatch (Kent)
Shipston on Stour (Warwickshire)
Shipton (Gloucestershire)
Shipton (North Yorkshire)
Shipton (Shropshire)
Shipton Moyne (Gloucestershire)
Shiptonthorpe (East Riding of Yorkshire)
Shirdley Hill (Lancashire)
Shire (Cumbria)
SHIRE OAK (West Midlands)
SHIREBROOK (Derbyshire)
SHIREGREEN (South Yorkshire)
SHIREHAMPTON (Bristol)
SHIREMOOR (Tyne & Wear)
Shirenewton (Monmouthshire)
SHIREOAKS (Nottinghamshire)
Shirkoak (Kent)
Shirl Heath (Hereford & Worcester)
SHIRLAND (Derbyshire)
Shirlett (Shropshire)
Shirley (Derbyshire)
Shirley (West Midlands)
Shittlehope (Durham)
Shobdon (Hereford & Worcester)
SHOBY (Leicestershire)
Shocklach (Cheshire)
Shocklach Green (Cheshire)
SHOLDEN (Kent)
SHOOSE (Cumbria)
Shoot Hill (Shropshire)
SHORE (Greater Manchester)
Shoreditch (Somerset)
Shoreham (Kent)
SHORESWOOD (Northumberland)
Shorncote (Gloucestershire)
Shorne (Kent)
Shorne Ridgeway (Kent)
SHORT HEATH (West Midlands)

SHOSCOMBE (Somerset)
SHOTLEY BRIDGE (Northumberland)
SHOTLEYFIELD (Northumberland)
Shottenden (Kent)
Shottery (Warwickshire)
Shotteswell (Warwickshire)
Shottle (Derbyshire)
Shottlegate (Derbyshire)
Shotton (Durham)
SHOTTON (Flintshire)
SHOTTON (Northumberland)
SHOTTON COLLIERY (Durham)
Shotwick (Cheshire)
Shoulton (Hereford & Worcester)
SHRALEYBROOK (Staffordshire)
Shrawardine (Shropshire)
Shrawley (Hereford & Worcester)
Shrewley (Warwickshire)
SHREWSBURY (Shropshire)
Shucknall (Hereford & Worcester)
Shurdington (Gloucestershire)
Shurnock (Hereford & Worcester)
Shurton (Somerset)
SHUSTOKE (Warwickshire)
SHUT END (West Midlands)
SHUT HEATH (Staffordshire)
Shuthonger (Gloucestershire)
Shutt Green (Staffordshire)
SHUTTINGTON (Warwickshire)
SHUTTLEWOOD (Derbyshire)
SHUTTLEWOOD COMMON (Derbyshire)
SHUTTLEWORTH (Lancashire)
Sibdon Carwood (Shropshire)
Sibsey (Lincolnshire)
Sibsey Fenside (Lincolnshire)
Sibson (Leicestershire)
Sibthorpe (Nottinghamshire)
Sicklinghall (North Yorkshire)
SID COP (South Yorkshire)
Sidbrook (Somerset)
SIDBURY (Shropshire)
Sidcot (Somerset)
SIDDICK (Cumbria)
Siddington (Cheshire)
Siddington (Gloucestershire)
Sidemoor (Hereford & Worcester)
Siefton (Shropshire)
Sigglesthorne (East Riding of Yorkshire)
Sigingstone (Vale of Glamorgan)
Sileby (Leicestershire)
Silecroft (Cumbria)
Silk Willoughby (Lincolnshire)
SILKSTONE (South Yorkshire)
SILKSTONE COMMON (South Yorkshire)
SILKSWORTH (Tyne & Wear)
Silloth (Cumbria)
Silpho (North Yorkshire)
SILSDEN (West Yorkshire)
Silver Street (Hereford & Worcester)
Silver Street (Kent)
Silver Street (Somerset)
Silverdale (Lancashire)
SILVERDALE (Staffordshire)
Silverdale Green (Lancashire)
SILVINGTON (Shropshire)
SIMMONDLEY (Derbyshire)
Simonburn (Northumberland)
Simonsbath (Somerset)
SIMONSTONE (Lancashire)
SIMONSTONE (North Yorkshire)
SIMPSON CROSS (Pembrokeshire)
Sinderby (North Yorkshire)
Sinderhope (Northumberland)
Sinderland Green (Greater Manchester)
Singleton (Lancashire)
Singlewell (Kent)
Sinkhurst Green (Kent)
Sinnington (North Yorkshire)
Sinton (Hereford & Worcester)
Sinton Green (Hereford & Worcester)
SIRHOWY (Blaenau Gwent)
Sissinghurst (Kent)
SISTON (Gloucestershire)
Sittingbourne (Kent)
Six Ashes (Staffordshire)
SIX BELLS (Blaenau Gwent)
Six Mile Cottages (Kent)
Sixhills (Lincolnshire)
Skeeby (North Yorkshire)
Skeffington (Leicestershire)
Skeffling (East Riding of Yorkshire)

See paras. 2 and 4 of the User Guide 2003 if you can't find your place name.

SKEGBY (Nottinghamshire)
Skegness (Lincolnshire)
SKELBROOKE (South Yorkshire)
Skeldyke (Lincolnshire)
Skellingthorpe (Lincolnshire)
SKELLORM GREEN (Cheshire)
SKELLOW (South Yorkshire)
SKELMANTHORPE (West Yorkshire)
SKELMERSDALE (Lancashire)
Skelton (Cumbria)
Skelton (East Riding of Yorkshire)
SKELTON (North Yorkshire)
Skelwith Bridge (Cumbria)
Skendleby (Lincolnshire)
Skenfrith (Monmouthshire)
Skerne (East Riding of Yorkshire)
Skerton (Lancashire)
Sketchley (Leicestershire)
SKETTY (Swansea)
SKEWEN (Neath Port Talbot)
SKEWSBY (North Yorkshire)
Skidbrooke (Lincolnshire)
Skidbrooke North End (Lincolnshire)
Skidby (East Riding of Yorkshire)
Skilgate (Somerset)
Skillington (Lincolnshire)
Skinburness (Cumbria)
Skinningrove (North Yorkshire)
SKIPRIGG (Cumbria)
Skipsea (East Riding of Yorkshire)
Skipsea Brough (East Riding of Yorkshire)
SKIPTON (North Yorkshire)
Skipton-on-Swale (North Yorkshire)
SKIPWITH (North Yorkshire)
Skirpenbeck (East Riding of Yorkshire)
Skirwith (Cumbria)
SKIRWITH (North Yorkshire)
Skitby (Cumbria)
Skyborry Green (Shropshire)
SKYREHOLME (North Yorkshire)
Slack (Derbyshire)
Slack (West Yorkshire)
Slack Head (Cumbria)
SLACK SIDE (West Yorkshire)
SLACKCOTE (Greater Manchester)
Slackholme End (Lincolnshire)
Slad (Gloucestershire)
Slade (Somerset)
Slade Green (Kent)
SLADE HEATH (Staffordshire)
SLADE HOOTON (South Yorkshire)
SLADEN (Derbyshire)
Slades Green (Hereford & Worcester)
SLAGGYFORD (Northumberland)
Slaid Hill (West Yorkshire)
Slaidburn (Lancashire)
Slaithwaite (West Yorkshire)
Slaley (Derbyshire)
Slaley (Northumberland)
SLATTOCKS (Greater Manchester)
Slawston (Leicestershire)
Sleaford (Lincolnshire)
SLEAGILL (Cumbria)
Sleap (Shropshire)
Sleapford (Shropshire)
Sledge Green (Hereford & Worcester)
Sledmere (East Riding of Yorkshire)
Sleightholme (Durham)
Sleights (North Yorkshire)
Slimbridge (Gloucestershire)
Slindon (Staffordshire)
Slingsby (North Yorkshire)
SLITTING MILL (Staffordshire)
Sloothby (Lincolnshire)
Slough Green (Somerset)
Slyne (Lancashire)
Small Heath (West Midlands)
Small Hythe (Kent)
Small Wood Hey (Lancashire)
SMALLBRIDGE (Greater Manchester)
Smallbrook (Gloucestershire)
Smalldale (Derbyshire)
SMALLEY (Derbyshire)
SMALLEY COMMON (Derbyshire)
SMALLEY GREEN (Derbyshire)
SMALLTHORNE (Staffordshire)
Smallwood (Cheshire)
SMARDALE (Cumbria)
Smarden (Kent)
Smarden Bell (Kent)
Smart's Hill (Kent)

SMEAFIELD (Northumberland)
Smeeth (Kent)
Smeeton Westerby (Leicestershire)
SMELTHOUSES (North Yorkshire)
Smestow (Staffordshire)
SMETHWICK (West Midlands)
Smethwick Green (Cheshire)
SMISBY (Derbyshire)
Smith End Green (Hereford & Worcester)
Smith Green (Lancashire)
Smithfield (Cumbria)
SMITHIES (South Yorkshire)
SMITHY BRIDGE (Greater Manchester)
Smithy Green (Cheshire)
Smithy Green (Greater Manchester)
SMITHY HOUSES (Derbyshire)
Smockington (Leicestershire)
Snailbeach (Shropshire)
Snainton (North Yorkshire)
SNAITH (East Riding of Yorkshire)
Snake Pass Inn (Derbyshire)
Snape (North Yorkshire)
Snape Green (Merseyside)
SNARESTONE (Leicestershire)
Snarford (Lincolnshire)
Snargate (Kent)
Snave (Kent)
Sneachill (Hereford & Worcester)
Snead (Powys)
Sneaton (North Yorkshire)
SNEATON HIGH MOOR (North Yorkshire)
SNEATONTHORPE (North Yorkshire)
Snelland (Lincolnshire)
Snelston (Derbyshire)
Sneyd Park (Bristol)
SNIBSTON (Leicestershire)
Snig's End (Gloucestershire)
Snitter (Northumberland)
Snitterby (Lincolnshire)
Snitterfield (Warwickshire)
Snitterton (Derbyshire)
SNITTLEGARTH (Cumbria)
SNITTON (Shropshire)
Snoadhill (Kent)
Snodhill (Hereford & Worcester)
Snodland (Kent)
Snoll Hatch (Kent)
SNOWDEN HILL (South Yorkshire)
Snowshill (Gloucestershire)
SOAR (Cardiff)
Soar (Powys)
Sockbridge (Cumbria)
Sockburn (Durham)
Sodom (Denbighshire)
SODYLT BANK (Shropshire)
SOLBURY (Pembrokeshire)
Sole Street (Kent)
SOLIHULL (West Midlands)
Sollers Dilwyn (Hereford & Worcester)
Sollers Hope (Hereford & Worcester)
Sollom (Lancashire)
Solva (Pembrokeshire)
Somerby (Leicestershire)
Somerby (Lincolnshire)
SOMERCOTES (Derbyshire)
Somerford Keynes (Gloucestershire)
Somersal Herbert (Derbyshire)
Somersby (Lincolnshire)
Somerton (Somerset)
Somerwood (Shropshire)
SONTLEY (Wrexham)
Sosgill (Cumbria)
Sotby (Lincolnshire)
Sots Hole (Lincolnshire)
SOUGHTON (Flintshire)
Soulby (Cumbria)
SOUNDWELL (Gloucestershire)
Soutergate (Cumbria)
South Alkham (Kent)
SOUTH ANSTON (South Yorkshire)
South Ashford (Kent)
South Bank (North Yorkshire)
South Barrow (Somerset)
SOUTH BRAMWITH (South Yorkshire)
South Brewham (Somerset)
SOUTH BROOMHILL (Northumberland)
South Cadbury (Somerset)
South Carlton (Lincolnshire)
SOUTH CARLTON (Nottinghamshire)
South Cave (East Riding of Yorkshire)
South Cerney (Gloucestershire)

South Chard (Somerset)
SOUTH CHARLTON (Northumberland)
South Cheriton (Somerset)
SOUTH CHURCH (Durham)
South Cleatlam (Durham)
South Cliffe (East Riding of Yorkshire)
SOUTH CLIFTON (Nottinghamshire)
South Cockerington (Lincolnshire)
South Collingham (Nottinghamshire)
South Cornelly (Bridgend)
SOUTH CROSLAND (West Yorkshire)
South Croxton (Leicestershire)
South Dalton (East Riding of Yorkshire)
South Darenth (Kent)
SOUTH DUFFIELD (North Yorkshire)
South Elkington (Lincolnshire)
SOUTH ELMSALL (West Yorkshire)
South End (East Riding of Yorkshire)
South End (Hereford & Worcester)
South End (Lincolnshire)
South Ferriby (Lincolnshire)
South Field (East Riding of Yorkshire)
SOUTH GOSFORTH (Tyne & Wear)
SOUTH GOSWORTH (Tyne & Wear)
South Green (Kent)
South Hazelrigg (Northumberland)
SOUTH HETTON (Durham)
SOUTH HIENDLEY (West Yorkshire)
South Hill (Somerset)
South Hykeham (Lincolnshire)
SOUTH HYLTON (Tyne & Wear)
South Kelsey (Lincolnshire)
South Killingholme (Lincolnshire)
South Kilvington (North Yorkshire)
SOUTH KIRKBY (West Yorkshire)
South Kyme (Lincolnshire)
South Leverton (Nottinghamshire)
South Littleton (Hereford & Worcester)
South Middleton (Northumberland)
SOUTH MILFORD (North Yorkshire)
SOUTH MOOR (Durham)
SOUTH MUSKHAM (Nottinghamshire)
South Newbald (East Riding of Yorkshire)
SOUTH NORMANTON (Derbyshire)
South Ormsby (Lincolnshire)
SOUTH OSSETT (West Yorkshire)
South Otterington (North Yorkshire)
South Owersby (Lincolnshire)
South Petherton (Somerset)
South Quarme (Somerset)
South Rauceby (Lincolnshire)
SOUTH REDDISH (Greater Manchester)
South Reston (Lincolnshire)
SOUTH SCARLE (Nottinghamshire)
SOUTH SHIELDS (Tyne & Wear)
South Shore (Lancashire)
South Somercotes (Lincolnshire)
South Stainley (North Yorkshire)
South Stoke (Somerset)
South Stour (Kent)
South Street (Kent)
South Thoresby (Lincolnshire)
South Thorpe (Durham)
SOUTH WIDCOMBE (Somerset)
South Wigston (Leicestershire)
South Willesborough (Kent)
South Willingham (Lincolnshire)
SOUTH WINGATE (Durham)
SOUTH WINGFIELD (Derbyshire)
South Witham (Lincolnshire)
Southam (Gloucestershire)
Southam (Warwickshire)
Southborough (Kent)
Southburn (East Riding of Yorkshire)
SOUTHERNBY (Cumbria)
Southernden (Kent)
Southerndown (Vale of Glamorgan)
Southfleet (Kent)
Southgate (Swansea)
Southmead (Bristol)
SOUTHOWRAM (West Yorkshire)
Southport (Merseyside)
Southrey (Lincolnshire)
Southrop (Gloucestershire)
SOUTHSEA (Wrexham)
SOUTHSIDE (Durham)
Southtown (Somerset)
SOUTHVILLE (Bristol)
Southwaite (Cumbria)
Southway (Somerset)
Southwell (Nottinghamshire)

See paras. 2 and 4 of the User Guide 2003 if you can't find your place name.

Southwick (Somerset)
SOUTHWICK (Tyne & Wear)
Southwood (Somerset)
SOWE COMMON (West Midlands)
Sower Carr (Lancashire)
Sowerby (North Yorkshire)
Sowerby (West Yorkshire)
Sowerby Bridge (West Yorkshire)
SOWERBY ROW (Cumbria)
Sowerhill (Somerset)
Sowhill (Torfaen)
SOWOOD (West Yorkshire)
Soyland Town (West Yorkshire)
Spalding (Lincolnshire)
Spaldington (East Riding of Yorkshire)
SPALFORD (Nottinghamshire)
Spanby (Lincolnshire)
Spark Bridge (Cumbria)
Sparket (Cumbria)
Sparkford (Somerset)
Sparkhill (West Midlands)
Sparrowpit (Derbyshire)
SPARTYLEA (Cumbria)
Spath (Staffordshire)
SPAUNTON (North Yorkshire)
Spaxton (Somerset)
SPEEDWELL (Bristol)
Speeton (North Yorkshire)
Speke (Merseyside)
Speldhurst (Kent)
Spelmonden (Kent)
SPEN (West Yorkshire)
Spen Green (Cheshire)
Spennithorne (North Yorkshire)
SPENNYMOOR (Durham)
Spernall (Warwickshire)
Spetchley (Hereford & Worcester)
Spilsby (Lincolnshire)
SPINDLESTONE (Northumberland)
SPINKHILL (Derbyshire)
Spital (Merseyside)
SPITAL HILL (South Yorkshire)
Spital in the Street (Lincolnshire)
Spittal (East Riding of Yorkshire)
SPITTAL (Northumberland)
Spittal (Pembrokeshire)
Splottlands (Cardiff)
Spodegreen (Cheshire)
Spofforth (North Yorkshire)
SPON GREEN (Flintshire)
Spondon (Derbyshire)
Spridlington (Lincolnshire)
SPRING VALE (South Yorkshire)
SPRINGHILL (Staffordshire)
Springthorpe (Lincolnshire)
SPRINGWELL (Tyne & Wear)
Sproatley (East Riding of Yorkshire)
Sproston Green (Cheshire)
SPROTBROUGH (South Yorkshire)
Sproxton (Leicestershire)
Sproxton (North Yorkshire)
Spunhill (Shropshire)
Spurstow (Cheshire)
St Andrew's Major (Vale of Glamorgan)
ST ANNE'S (Bristol)
St Anne's (Lancashire)
ST ANNE'S PARK (Bristol)
St Arvans (Monmouthshire)
St Asaph (Denbighshire)
St Athan (Vale of Glamorgan)
ST BEES (Cumbria)
St Briavels (Gloucestershire)
St Bride's Major (Vale of Glamorgan)
St Brides (Pembrokeshire)
St Brides Netherwent (Monmouthshire)
St Brides super-Ely (Cardiff)
St Brides Wentlooge (Newport)
St Catherine (Somerset)
St Chloe (Gloucestershire)
St Clears (Carmarthenshire)
St Davids (Pembrokeshire)
St Decumans (Somerset)
St Devereux (Hereford & Worcester)
St Dogmaels (Cardiganshire)
St Dogwells (Pembrokeshire)
St Donats (Vale of Glamorgan)
St Fagans (Cardiff)
St Florence (Pembrokeshire)
ST GEORGE (Bristol)
St George's (Vale of Glamorgan)
St Georges (Somerset)

St Harmon (Powys)
ST HELEN AUCKLAND (Durham)
ST HELENS (Cumbria)
ST HELENS (Merseyside)
St Hilary (Vale of Glamorgan)
ST ILLTYD (Blaenau Gwent)
St Ishmaels (Pembrokeshire)
St John's Chapel (Durham)
St Johns (Durham)
St Johns (Hereford & Worcester)
St Johns (Kent)
St Laurence (Kent)
St Leonard's Street (Kent)
St Lythans (Vale of Glamorgan)
St Margaret's at Cliffe (Kent)
St Margarets (Hereford & Worcester)
ST MARTIN'S MOOR (Shropshire)
ST MARTINS (Shropshire)
St Mary Church (Vale of Glamorgan)
St Mary Hill (Vale of Glamorgan)
St Mary in the Marsh (Kent)
St Mary's Bay (Kent)
ST MARY'S GROVE (Somerset)
St Mary's Hoo (Kent)
St Maughans (Monmouthshire)
St Maughans Green (Monmouthshire)
St Mellons (Cardiff)
St Michael Church (Somerset)
St Michael's on Wyre (Lancashire)
St Michaels (Hereford & Worcester)
St Michaels (Kent)
St Nicholas (Pembrokeshire)
St Nicholas (Vale of Glamorgan)
St Nicholas at Wade (Kent)
St Owens Cross (Hereford & Worcester)
ST PAUL'S (Bristol)
St Peter's (Kent)
St Petrox (Pembrokeshire)
ST PHILIP'S MARSH (Bristol)
St Twynnells (Pembrokeshire)
St Weonards (Hereford & Worcester)
St-y-Nyll (Vale of Glamorgan)
Stableford (Shropshire)
STABLEFORD (Staffordshire)
STACEY BANK (Derbyshire)
Stackhouse (North Yorkshire)
Stackpole (Pembrokeshire)
STACKSTEADS (Lancashire)
Staddlethorpe (East Riding of Yorkshire)
Staffield (Cumbria)
Stafford (Staffordshire)
Stag Green (Cumbria)
STAINBURN (Cumbria)
STAINBURN (North Yorkshire)
Stainby (Lincolnshire)
STAINCROSS (South Yorkshire)
Staindrop (Durham)
Stainfield (Lincolnshire)
Stainforth (North Yorkshire)
STAINFORTH (South Yorkshire)
Staining (Lancashire)
STAINLAND (West Yorkshire)
STAINSACRE (North Yorkshire)
STAINSBY (Derbyshire)
Stainton (Cumbria)
Stainton (Durham)
Stainton (North Yorkshire)
STAINTON (South Yorkshire)
Stainton by Langworth (Lincolnshire)
Stainton le Vale (Lincolnshire)
Stainton with Adgarley (Cumbria)
Staintondale (North Yorkshire)
Stair (Cumbria)
STAIRFOOT (South Yorkshire)
Staithes (North Yorkshire)
Stake Pool (Lancashire)
STAKEFORD (Northumberland)
Stalisfield Green (Kent)
STALLING BUSK (North Yorkshire)
Stallingborough (Lincolnshire)
STALLINGTON (Staffordshire)
Stalmine (Lancashire)
Stalmine Moss Side (Lancashire)
STALYBRIDGE (Greater Manchester)
Stamford (Lincolnshire)
STAMFORD (Northumberland)
Stamford Bridge (Cheshire)
Stamford Bridge (East Riding of Yorkshire)
STAMFORDHAM (Northumberland)
Stamton Lees (Derbyshire)
Stanah (Lancashire)

STANBURY (West Yorkshire)
STAND (Greater Manchester)
Standeford (Staffordshire)
Standen (Kent)
Standen Street (Kent)
Standerwick (Somerset)
STANDINGSTONE (Cumbria)
STANDISH (Gloucestershire)
STANDISH (Greater Manchester)
STANDISH LOWER GROUND (Greater Manchester)
STANDON (Staffordshire)
Stanford (Kent)
STANFORD (Shropshire)
Stanford Bishop (Hereford & Worcester)
Stanford Bridge (Hereford & Worcester)
Stanford Bridge (Shropshire)
Stanford on Soar (Nottinghamshire)
Stanford on Teme (Hereford & Worcester)
STANFREE (Derbyshire)
Stanghow (North Yorkshire)
STANHILL (Lancashire)
STANHOPE (Durham)
STANHOPE BRETBY (Derbyshire)
STANLEY (Derbyshire)
STANLEY (Durham)
STANLEY (Nottinghamshire)
STANLEY (Shropshire)
STANLEY (Staffordshire)
STANLEY (West Yorkshire)
STANLEY COMMON (Derbyshire)
STANLEY CROOK (Durham)
STANLEY FERRY (West Yorkshire)
STANLEY GATE (Lancashire)
STANLEY MOOR (Staffordshire)
Stanley Pontlarge (Gloucestershire)
Stannersburn (Northumberland)
STANNINGLEY (West Yorkshire)
STANNINGTON (Northumberland)
STANNINGTON (South Yorkshire)
STANNINGTON STATION (Northumberland)
Stansbatch (Hereford & Worcester)
Stanshope (Staffordshire)
Stansted (Kent)
STANTON (Derbyshire)
Stanton (Gloucestershire)
Stanton (Monmouthshire)
STANTON (Northumberland)
Stanton (Staffordshire)
Stanton by Bridge (Derbyshire)
STANTON BY DALE (Derbyshire)
STANTON DREW (Somerset)
STANTON HILL (Nottinghamshire)
Stanton in Peak (Derbyshire)
Stanton Lacy (Shropshire)
Stanton Long (Shropshire)
STANTON ON THE WOLDS (Nottinghamshire)
Stanton Prior (Somerset)
STANTON UNDER BARDON (Leicestershire)
Stanton upon Hine Heath (Shropshire)
STANTON WICK (Somerset)
Stantway (Gloucestershire)
Stanwardine in the Field (Shropshire)
Stanwardine in the Wood (Shropshire)
Stanway (Gloucestershire)
Stanwix (Cumbria)
STAPE (North Yorkshire)
Stapeley (Cheshire)
STAPENHILL (Staffordshire)
STAPLE (Kent)
Staple (Somerset)
Staple Fitzpaine (Somerset)
Staple Hill (Hereford & Worcester)
Stapleford (Leicestershire)
STAPLEFORD (Lincolnshire)
STAPLEFORD (Nottinghamshire)
Staplegrove (Somerset)
Staplehay (Somerset)
Staplehurst (Kent)
Staplestreet (Kent)
STAPLET (Cumbria)
STAPLETON (Bristol)
Stapleton (Hereford & Worcester)
Stapleton (Leicestershire)
Stapleton (North Yorkshire)
STAPLETON (Shropshire)
Stapleton (Somerset)
Stapley (Somerset)
Staplow (Hereford & Worcester)
Star (Pembrokeshire)
Star (Somerset)
Starbeck (North Yorkshire)

STARBOTTON (North Yorkshire)
Stareton (Warwickshire)
Starkholmes (Derbyshire)
Starklin (Hereford & Worcester)
STARLING (Greater Manchester)
Startforth (Durham)
STATENBOROUGH (Kent)
Statham (Cheshire)
Stathe (Somerset)
STATHERN (Leicestershire)
STATION TOWN (Durham)
STAUNTON (Gloucestershire)
Staunton Green (Hereford & Worcester)
Staunton in the Vale (Nottinghamshire)
Staunton on Arrow (Hereford & Worcester)
Staunton on Wye (Hereford & Worcester)
Staveley (Cumbria)
STAVELEY (Derbyshire)
Staveley (North Yorkshire)
Staverton (Gloucestershire)
Staverton Bridge (Gloucestershire)
Stawell (Somerset)
Stawley (Somerset)
Staxton (North Yorkshire)
Staylittle (Powys)
Staynall (Lancashire)
Staythorpe (Nottinghamshire)
Stead (West Yorkshire)
STEAN (North Yorkshire)
STEARSBY (North Yorkshire)
Steart (Somerset)
Stechford (West Midlands)
Stede Quarter (Kent)
Steel (Northumberland)
Steel Heath (Shropshire)
Steen's Bridge (Hereford & Worcester)
Steep Lane (West Yorkshire)
Steeton (West Yorkshire)
STELLA (Tyne & Wear)
Stelling Minnis (Kent)
Stembridge (Somerset)
Stenigot (Lincolnshire)
STEPASIDE (Pembrokeshire)
STEPPING HILL (Greater Manchester)
Stewley (Somerset)
Stewton (Lincolnshire)
Steynton (Pembrokeshire)
Stickford (Lincolnshire)
Sticklepath (Somerset)
Stickney (Lincolnshire)
Stidd (Lancashire)
Stiff Green (Kent)
Stifford's Bridge (Hereford & Worcester)
Stile Bridge (Kent)
Stileway (Somerset)
STILLINGFLEET (North Yorkshire)
Stillington (Durham)
Stillington (North Yorkshire)
Stinchcombe (Gloucestershire)
Stiperstones (Shropshire)
STIRCHLEY (Shropshire)
Stirchley (West Midlands)
Stirton (North Yorkshire)
STIVICHALL (West Midlands)
Stixwould (Lincolnshire)
Stoak (Cheshire)
STOBSWOOD (Northumberland)
Stock (Somerset)
Stock Green (Hereford & Worcester)
Stock Wood (Hereford & Worcester)
Stockbury (Kent)
STOCKDALEWATH (Cumbria)
Stocker's Hill (Kent)
Stockerston (Leicestershire)
Stocking (Hereford & Worcester)
STOCKINGFORD (Warwickshire)
Stockland Bristol (Somerset)
Stockland Green (Kent)
Stockley Hill (Hereford & Worcester)
Stocklinch (Somerset)
Stockmoor (Hereford & Worcester)
STOCKPORT (Greater Manchester)
STOCKSBRIDGE (South Yorkshire)
STOCKSFIELD (Northumberland)
Stockton (Hereford & Worcester)
STOCKTON (Shropshire)
Stockton (Warwickshire)
STOCKTON BROOK (Staffordshire)
Stockton Heath (Cheshire)
STOCKTON ON TEME (Hereford & Worcester)
Stockton on the Forest (North Yorkshire)

Stockton-on-Tees (Durham)
Stockwell (Gloucestershire)
STOCKWELL END (West Midlands)
Stockwell Heath (Staffordshire)
STOCKWOOD (Bristol)
Stodday (Lancashire)
STODMARSH (Kent)
Stoford (Somerset)
Stogumber (Somerset)
Stogursey (Somerset)
Stoke (Kent)
STOKE (West Midlands)
STOKE BARDOLPH (Nottinghamshire)
Stoke Bishop (Bristol)
Stoke Bliss (Hereford & Worcester)
Stoke Cross (Hereford & Worcester)
Stoke Edith (Hereford & Worcester)
Stoke End (Warwickshire)
STOKE GIFFORD (Gloucestershire)
Stoke Golding (Leicestershire)
Stoke Heath (Hereford & Worcester)
Stoke Heath (Shropshire)
STOKE HEATH (West Midlands)
Stoke Lacy (Hereford & Worcester)
Stoke Orchard (Gloucestershire)
Stoke Pound (Hereford & Worcester)
Stoke Prior (Hereford & Worcester)
Stoke Rochford (Lincolnshire)
Stoke St. Gregory (Somerset)
Stoke St. Mary (Somerset)
Stoke St. Michael (Somerset)
STOKE ST. MILBOROUGH (Shropshire)
Stoke sub Hamdon (Somerset)
Stoke Trister (Somerset)
Stoke upon Tern (Shropshire)
Stoke Wharf (Hereford & Worcester)
STOKE-ON-TRENT (Staffordshire)
Stokeham (Nottinghamshire)
Stokesay (Shropshire)
Stokesley (North Yorkshire)
Stolford (Somerset)
STON EASTON (Somerset)
Stone (Gloucestershire)
Stone (Hereford & Worcester)
Stone (Kent)
Stone (Somerset)
STONE (South Yorkshire)
Stone (Staffordshire)
Stone Allerton (Somerset)
STONE CHAIR (West Yorkshire)
STONE CROSS (Kent)
STONE HILL (South Yorkshire)
Stone House (Cumbria)
STONE ROWS (Leicestershire)
Stone Street (Kent)
STONE-EDGE-BATCH (Somerset)
Stonebridge (Somerset)
STONEBRIDGE (West Midlands)
STONEBROOM (Derbyshire)
Stonecrouch (Kent)
Stoneferry (East Riding of Yorkshire)
STONEGARTHSIDE (Cumbria)
STONEGATE (North Yorkshire)
Stonegrave (North Yorkshire)
Stonehall (Hereford & Worcester)
STONEHAUGH (Northumberland)
Stonehouse (Cheshire)
Stonehouse (Gloucestershire)
STONEHOUSE (Northumberland)
Stoneleigh (Warwickshire)
Stoneley Green (Cheshire)
Stonesby (Leicestershire)
Stonethwaite (Cumbria)
Stonetree Green (Kent)
Stonewood (Kent)
Stoney Middleton (Derbyshire)
Stoney Stanton (Leicestershire)
Stoney Stoke (Somerset)
Stoney Stratton (Somerset)
STONEY STRETTON (Shropshire)
Stoneybridge (Hereford & Worcester)
Stoneygate (Leicestershire)
STONNALL (Staffordshire)
Stonton Wyville (Leicestershire)
Stony Cross (Hereford & Worcester)
STONY HOUGHTON (Derbyshire)
STONYWELL (Staffordshire)
Storeton (Merseyside)
Storeyard Green (Hereford & Worcester)
STORMY CORNER (Lancashire)
Storridge (Hereford & Worcester)

Storth (Cumbria)
STORWOOD (East Riding of Yorkshire)
STOTTESDON (Shropshire)
Stoughton (Leicestershire)
Stoulton (Hereford & Worcester)
STOURBRIDGE (West Midlands)
STOURPORT-ON-SEVERN (Hereford & Worcester)
Stourton (Staffordshire)
Stourton (Warwickshire)
STOURTON (West Yorkshire)
Stout (Somerset)
Stow (Lincolnshire)
Stow-on-the-Wold (Gloucestershire)
STOWE (Gloucestershire)
Stowe (Shropshire)
Stowe by Chartley (Staffordshire)
Stowell (Somerset)
STOWEY (Somerset)
Stowting (Kent)
Stowting Common (Kent)
Stragglethorpe (Lincolnshire)
STRAGGLETHORPE (Nottinghamshire)
Stramshall (Staffordshire)
Strangford (Hereford & Worcester)
Stratford-upon-Avon (Warwickshire)
Stratton (Gloucestershire)
STRATTON-ON-THE-FOSSE (Somerset)
Stream (Somerset)
Street (Lancashire)
STREET (North Yorkshire)
Street (Somerset)
Street Ashton (Warwickshire)
STREET DINAS (Shropshire)
Street End (Kent)
STREET GATE (Tyne & Wear)
Street Houses (North Yorkshire)
STREET LANE (Derbyshire)
Street on the Fosse (Somerset)
Streethay (Staffordshire)
Streetlam (North Yorkshire)
STREETLY (West Midlands)
Strefford (Shropshire)
STRELLEY (Nottinghamshire)
Strensall (North Yorkshire)
Strensham (Hereford & Worcester)
Stretcholt (Somerset)
STRETFORD (Greater Manchester)
Stretford (Hereford & Worcester)
Stretton (Cheshire)
STRETTON (Derbyshire)
Stretton (Staffordshire)
STRETTON EN LE FIELD (Leicestershire)
Stretton Grandison (Hereford & Worcester)
STRETTON HEATH (Shropshire)
Stretton on Fosse (Warwickshire)
Stretton Sugwas (Hereford & Worcester)
Stretton under Fosse (Warwickshire)
Stretton Westwood (Shropshire)
Stretton-on-Dunsmore (Warwickshire)
STRINES (Greater Manchester)
Stringston (Somerset)
Stroat (Gloucestershire)
Strood (Kent)
Stroud (Gloucestershire)
Stroud Green (Gloucestershire)
Stroxton (Lincolnshire)
Strubby (Lincolnshire)
Stryd-y-Facsen (Anglesey)
STRYT-ISSA (Wrexham)
STUBBERS GREEN (West Midlands)
STUBBINS (North Yorkshire)
STUBLEY (Derbyshire)
STUBSHAW CROSS (Greater Manchester)
Stubton (Lincolnshire)
Studfold (North Yorkshire)
Studholme (Cumbria)
Studley (Warwickshire)
Studley Common (Hereford & Worcester)
Studley Roger (North Yorkshire)
Studley Royal (North Yorkshire)
Sturbridge (Staffordshire)
Sturgate (Lincolnshire)
Sturry (Kent)
Sturton (Lincolnshire)
Sturton by Stow (Lincolnshire)
Sturton le Steeple (Nottinghamshire)
Stutton (North Yorkshire)
Styal (Cheshire)
STYRRUP (Nottinghamshire)
Suckley (Hereford & Worcester)
Suckley Green (Hereford & Worcester)

See paras. 2 and 4 of the User Guide 2003 if you can't find your place name.

Sudbrook (Lincolnshire)
Sudbrook (Monmouthshire)
Sudbrooke (Lincolnshire)
Sudbury (Derbyshire)
SUDDEN (Greater Manchester)
Suddington (Hereford & Worcester)
Sudgrove (Gloucestershire)
Suffield (North Yorkshire)
Sugdon (Shropshire)
Sugnall (Staffordshire)
Sugwas Pool (Hereford & Worcester)
Sully (Vale of Glamorgan)
SUMMER HILL (Wrexham)
SUMMERBRIDGE (North Yorkshire)
Summerfield (Hereford & Worcester)
Summerhouse (Durham)
Summerlands (Cumbria)
SUMMERLEY (Derbyshire)
SUMMERSEAT (Greater Manchester)
SUMMIT (Greater Manchester)
SUNBIGGIN (North Yorkshire)
Sunderland (Cumbria)
Sunderland (Lancashire)
SUNDERLAND (Tyne & Wear)
SUNDERLAND BRIDGE (Durham)
Sundridge (Kent)
Sunk Island (East Riding of Yorkshire)
SUNNISIDE (Durham)
SUNNISIDE (Tyne & Wear)
SUNNY BANK (Lancashire)
SUNNY BROW (Durham)
Sunnyhill (Derbyshire)
SUNNYHURST (Lancashire)
Surfleet (Lincolnshire)
Surfleet Seas End (Lincolnshire)
Susworth (Lincolnshire)
Sutterby (Lincolnshire)
Sutterton (Lincolnshire)
SUTTON (Kent)
SUTTON (Merseyside)
Sutton (North Yorkshire)
SUTTON (Nottinghamshire)
Sutton (Pembrokeshire)
SUTTON (Shropshire)
SUTTON (South Yorkshire)
Sutton (Staffordshire)
Sutton at Hone (Kent)
Sutton Bingham (Somerset)
Sutton Bonington (Nottinghamshire)
Sutton Bridge (Lincolnshire)
Sutton Cheney (Leicestershire)
SUTTON COLDFIELD (West Midlands)
Sutton Crosses (Lincolnshire)
Sutton Grange (North Yorkshire)
SUTTON GREEN (Wrexham)
Sutton Howgrave (North Yorkshire)
SUTTON IN ASHFIELD (Nottinghamshire)
Sutton in the Elms (Leicestershire)
SUTTON LANE ENDS (Cheshire)
SUTTON MADDOCK (Shropshire)
Sutton Mallet (Somerset)
SUTTON MANOR (Merseyside)
Sutton Marsh (Hereford & Worcester)
Sutton Montis (Somerset)
Sutton on Sea (Lincolnshire)
Sutton on the Hill (Derbyshire)
SUTTON ON TRENT (Nottinghamshire)
SUTTON SCARSDALE (Derbyshire)
Sutton St. Edmund (Lincolnshire)
Sutton St. James (Lincolnshire)
Sutton St. Nicholas (Hereford & Worcester)
Sutton Street (Kent)
SUTTON UPON DERWENT (East Riding of Yorkshire)
Sutton Valence (Kent)
Sutton Weaver (Cheshire)
SUTTON WICK (Somerset)
SUTTON-IN-CRAVEN (North Yorkshire)
Sutton-on-Hull (East Riding of Yorkshire)
Sutton-on-the-Forest (North Yorkshire)
Sutton-under-Brailes (Warwickshire)
Sutton-under-Whitestonecliffe (North Yorkshire)
Swaby (Lincolnshire)
SWADLINCOTE (Derbyshire)
Swainby (North Yorkshire)
Swainshill (Hereford & Worcester)
Swainswick (Somerset)
Swalecliffe (Kent)
Swallow (Lincolnshire)
Swallow Beck (Lincolnshire)
SWALLOW NEST (South Yorkshire)
Swan Green (Cheshire)

SWAN VILLAGE (West Midlands)
Swanbridge (Vale of Glamorgan)
Swancote (Shropshire)
Swanland (East Riding of Yorkshire)
Swanley (Kent)
Swanley Village (Kent)
SWANNINGTON (Leicestershire)
Swanpool Garden Suberb (Lincolnshire)
Swanscombe (Kent)
SWANSEA (Swansea)
Swanton Street (Kent)
SWANWICK (Derbyshire)
Swarby (Lincolnshire)
Swarkestone (Derbyshire)
SWARLAND (Northumberland)
SWARLAND ESTATE (Northumberland)
Swartha (West Yorkshire)
Swarthmoor (Cumbria)
Swaton (Lincolnshire)
Swayfield (Lincolnshire)
Sweet Green (Hereford & Worcester)
Sweetlands Corner (Kent)
SWEPSTONE (Leicestershire)
Swettenham (Cheshire)
Swffryd (Blaenau Gwent)
Swift's Green (Kent)
Swillbrook (Lancashire)
SWILLINGTON (West Yorkshire)
SWINCLIFFE (North Yorkshire)
SWINCLIFFE (West Yorkshire)
Swinden (North Yorkshire)
SWINDERBY (Lincolnshire)
Swindon (Gloucestershire)
SWINDON (Northumberland)
Swindon (Staffordshire)
Swine (East Riding of Yorkshire)
SWINEFLEET (East Riding of Yorkshire)
SWINEFORD (Gloucestershire)
Swineshead (Lincolnshire)
Swineshead Bridge (Lincolnshire)
Swinford (Leicestershire)
Swingfield Minnis (Kent)
Swingfield Street (Kent)
SWINHOE (Northumberland)
Swinhope (Lincolnshire)
SWINITHWAITE (North Yorkshire)
Swinmore Common (Hereford & Worcester)
Swinscoe (Staffordshire)
Swinside (Cumbria)
Swinstead (Lincolnshire)
Swinthorpe (Lincolnshire)
SWINTON (Greater Manchester)
Swinton (North Yorkshire)
SWINTON (South Yorkshire)
Swithland (Leicestershire)
Sworton Heath (Cheshire)
SWYNNERTON (Staffordshire)
Sycharth (Powys)
Sychnant (Powys)
Sychtyn (Powys)
SYDALLT (Flintshire)
Syde (Gloucestershire)
Sydnal Lane (Shropshire)
Syerston (Nottinghamshire)
SYKE (Greater Manchester)
SYKEHOUSE (South Yorkshire)
SYLEN (Carmarthenshire)
Symonds Yat (Hereford & Worcester)
SYMPSON GREEN (West Yorkshire)
Syreford (Gloucestershire)
Syston (Leicestershire)
Syston (Lincolnshire)
Sytchampton (Hereford & Worcester)

T

Tabley Hill (Cheshire)
Tadcaster (North Yorkshire)
Taddington (Derbyshire)
Taddington (Gloucestershire)
Tadwick (Somerset)
Tafarn-y-bwlch (Pembrokeshire)
Tafarn-y-Gelyn (Denbighshire)
TAFARNAUBACH (Blaenau Gwent)
TAFF'S WELL (Cardiff)
Tafolwern (Powys)
Tai'r Bull (Powys)
TAIBACH (Neath Port Talbot)
Tal-y-coed (Monmouthshire)
Tal-y-garn (Rhondda Cynon Taff)
TAL-Y-WAUN (Torfaen)

Talachddu (Powys)
TALACRE (Flintshire)
Talbenny (Pembrokeshire)
TALBOT GREEN (Rhondda Cynon Taff)
Talerddig (Powys)
Talgarth (Powys)
TALKE (Staffordshire)
TALKE PITS (Staffordshire)
TALKIN (Cumbria)
TALLARN GREEN (Wrexham)
TALLENTIRE (Cumbria)
Talley (Carmarthenshire)
Tallington (Lincolnshire)
TALLWRN (Wrexham)
Talog (Carmarthenshire)
Talwrn (Anglesey)
TALWRN (Wrexham)
Talybont-on-Usk (Powys)
Talywern (Powys)
TAMER LANE END (Greater Manchester)
TAMWORTH (Staffordshire)
Tamworth Green (Lincolnshire)
TAN HILL (North Yorkshire)
TAN-Y-FRON (Wrexham)
Tancred (North Yorkshire)
Tancredston (Pembrokeshire)
TANFIELD (Durham)
TANFIELD LEA (Durham)
Tangiers (Pembrokeshire)
TANKERSLEY (South Yorkshire)
Tankerton (Kent)
Tanner Green (Hereford & Worcester)
Tansley (Derbyshire)
TANTOBIE (Durham)
Tanton (North Yorkshire)
Tanwood (Hereford & Worcester)
Tanworth in Arden (Warwickshire)
TARBOCK GREEN (Merseyside)
Tardebigge (Hereford & Worcester)
Tardy Gate (Lancashire)
Tarleton (Lancashire)
Tarlscough (Lancashire)
Tarlton (Gloucestershire)
Tarnock (Somerset)
Tarns (Cumbria)
Tarnside (Cumbria)
Tarporley (Cheshire)
Tarr (Somerset)
Tarrington (Hereford & Worcester)
Tarvin (Cheshire)
Tarvin Sands (Cheshire)
TASLEY (Shropshire)
TATENHILL (Staffordshire)
TATHAM (Lancashire)
Tathwell (Lincolnshire)
Tattenhall (Cheshire)
Tattershall (Lincolnshire)
Tattershall Bridge (Lincolnshire)
Tattershall Thorpe (Lincolnshire)
Tatworth (Somerset)
Taunton (Somerset)
Tavernspite (Pembrokeshire)
TAXAL (Derbyshire)
Taynton (Gloucestershire)
Tealby (Lincolnshire)
Tebay (Cumbria)
Teddington (Gloucestershire)
Tedstone Delamere (Hereford & Worcester)
Tedstone Wafer (Hereford & Worcester)
Teesport (North Yorkshire)
Tegryn (Pembrokeshire)
TELFORD (Shropshire)
Tellisford (Somerset)
Temple Balsall (West Midlands)
TEMPLE CLOUD (Somerset)
Temple Ewell (Kent)
Temple Grafton (Warwickshire)
Temple Guiting (Gloucestershire)
TEMPLE HIRST (North Yorkshire)
TEMPLE NORMANTON (Derbyshire)
Temple Sowerby (Cumbria)
Templecombe (Somerset)
TEMPLETON (Pembrokeshire)
TEMPLETOWN (Durham)
Tenbury Wells (Hereford & Worcester)
TENBY (Pembrokeshire)
Tenterden (Kent)
Tern (Shropshire)
Ternhill (Shropshire)
Terrington (North Yorkshire)
Terry's Green (Warwickshire)

See paras. 2 and 4 of the User Guide 2003 if you can't find your place name.

Teston (Kent)
Tetbury (Gloucestershire)
Tetbury Upton (Gloucestershire)
Tetchill (Shropshire)
Tetford (Lincolnshire)
Tetney (Lincolnshire)
Tetney Lock (Lincolnshire)
Tettenhall (West Midlands)
Tettenhall Wood (West Midlands)
TEVERSAL (Nottinghamshire)
Tewkesbury (Gloucestershire)
Teynham (Kent)
THACKLEY (West Yorkshire)
Thackthwaite (Cumbria)
Thanington (Kent)
THATTO HEATH (Merseyside)
THE BANK (Cheshire)
The Bank (Shropshire)
The Beeches (Gloucestershire)
The Blythe (Staffordshire)
The Bourne (Hereford & Worcester)
The Bratch (Staffordshire)
The Broad (Hereford & Worcester)
The Bush (Kent)
The Butts (Gloucestershire)
The Chequer (Wrexham)
The Corner (Kent)
The Corner (Shropshire)
The Flatt (Cumbria)
The Forge (Hereford & Worcester)
The Forstal (Kent)
The Fouralls (Shropshire)
The Green (Cumbria)
The Grove (Hereford & Worcester)
The Haw (Gloucestershire)
The Hill (Cumbria)
The Horns (Kent)
The Leacon (Kent)
THE MIDDLES (Durham)
The Moor (Kent)
The Mumbles (Swansea)
The Mythe (Gloucestershire)
The Narth (Monmouthshire)
The Quarry (Gloucestershire)
The Quarter (Kent)
The Reddings (Gloucestershire)
THE ROOKERY (Staffordshire)
THE SMITHIES (Shropshire)
The Spring (Warwickshire)
THE SQUARE (Torfaen)
The Stair (Kent)
The Stocks (Kent)
The Vauld (Hereford & Worcester)
THE WYKE (Shropshire)
Theakston (North Yorkshire)
Thealby (Lincolnshire)
Theale (Somerset)
Thearne (East Riding of Yorkshire)
Theddingworth (Leicestershire)
Theddlethorpe All Saints (Lincolnshire)
Theddlethorpe St. Helen (Lincolnshire)
THELWALL (Cheshire)
THETHWAITE (Cumbria)
THICKET PRIORY (East Riding of Yorkshire)
Thimbleby (Lincolnshire)
Thimbleby (North Yorkshire)
Thingwall (Merseyside)
THIRKLEBY (North Yorkshire)
Thirlby (North Yorkshire)
Thirlspot (Cumbria)
Thirn (North Yorkshire)
Thirsk (North Yorkshire)
Thirtleby (East Riding of Yorkshire)
Thistleton (Lancashire)
Thixendale (North Yorkshire)
Thockrington (Northumberland)
Tholthorpe (North Yorkshire)
THOMAS CHAPEL (Pembrokeshire)
Thomas Close (Cumbria)
Thomas Town (Warwickshire)
Thong (Kent)
THORALBY (North Yorkshire)
THORESBY (Nottinghamshire)
Thoresthorpe (Lincolnshire)
Thoresway (Lincolnshire)
Thorganby (Lincolnshire)
THORGANBY (North Yorkshire)
THORGILL (North Yorkshire)
Thorlby (North Yorkshire)
THORMANBY (North Yorkshire)
Thornaby-on-Tees (North Yorkshire)

Thornborough (North Yorkshire)
Thornbury (Gloucestershire)
Thornbury (Hereford & Worcester)
THORNBURY (West Yorkshire)
Thornby (Cumbria)
Thorncliff (Staffordshire)
Thorne (Somerset)
THORNE (South Yorkshire)
Thorne St. Margaret (Somerset)
Thorner (West Yorkshire)
THORNES (Staffordshire)
THORNES (West Yorkshire)
THORNEY (Nottinghamshire)
Thorney (Somerset)
Thornfalcon (Somerset)
THORNGRAFTON (Northumberland)
Thorngrove (Somerset)
Thorngumbald (East Riding of Yorkshire)
Thornhill (Caerphilly)
Thornhill (Derbyshire)
THORNHILL (West Yorkshire)
THORNHILL LEES (West Yorkshire)
THORNHILLS (West Yorkshire)
Thornholme (East Riding of Yorkshire)
Thornington (Northumberland)
THORNLEY (Durham)
Thornley Gate (Cumbria)
Thorns Green (Greater Manchester)
THORNSETT (Derbyshire)
Thornthwaite (Cumbria)
THORNTHWAITE (North Yorkshire)
Thornton (East Riding of Yorkshire)
Thornton (Lancashire)
THORNTON (Leicestershire)
Thornton (Lincolnshire)
Thornton (Merseyside)
Thornton (North Yorkshire)
THORNTON (Northumberland)
Thornton (Pembrokeshire)
THORNTON (West Yorkshire)
Thornton Curtis (Lincolnshire)
Thornton Dale (North Yorkshire)
Thornton Green (Cheshire)
Thornton Hough (Merseyside)
THORNTON IN LONSDALE (North Yorkshire)
Thornton le Moor (Lincolnshire)
THORNTON RUST (North Yorkshire)
Thornton Steward (North Yorkshire)
Thornton Watlass (North Yorkshire)
Thornton-in-Craven (North Yorkshire)
Thornton-le-Beans (North Yorkshire)
Thornton-le-Clay (North Yorkshire)
Thornton-le-Moor (North Yorkshire)
Thornton-le-Moors (Cheshire)
Thornton-le-Street (North Yorkshire)
Thornythwaite (Cumbria)
Thoroton (Nottinghamshire)
THORP ARCH (West Yorkshire)
Thorpe (Derbyshire)
Thorpe (East Riding of Yorkshire)
Thorpe (Lincolnshire)
THORPE (North Yorkshire)
Thorpe (Nottinghamshire)
Thorpe Acre (Leicestershire)
THORPE ARNOLD (Leicestershire)
THORPE AUDLIN (West Yorkshire)
Thorpe Bassett (North Yorkshire)
THORPE COMMON (South Yorkshire)
THORPE CONSTANTINE (Staffordshire)
THORPE GREEN (Lancashire)
THORPE HESLEY (South Yorkshire)
THORPE IN BALNE (South Yorkshire)
Thorpe in the Fallows (Lincolnshire)
Thorpe Langton (Leicestershire)
Thorpe Larches (Durham)
Thorpe le Street (East Riding of Yorkshire)
Thorpe on the Hill (Lincolnshire)
THORPE ON THE HILL (West Yorkshire)
THORPE SALVIN (South Yorkshire)
Thorpe Satchville (Leicestershire)
Thorpe St. Peter (Lincolnshire)
Thorpe Thewles (Durham)
Thorpe Tilney (Lincolnshire)
Thorpe Underwood (North Yorkshire)
THORPE WILLOUGHBY (North Yorkshire)
THREAPLAND (Cumbria)
THREAPLAND (North Yorkshire)
Threapwood (Cheshire)
THREAPWOOD (Staffordshire)
THREAPWOOD HEAD (Staffordshire)
Three Ashes (Hereford & Worcester)

Three Chimneys (Kent)
Three Cocks (Powys)
THREE CROSSES (Swansea)
Three Gates (Hereford & Worcester)
THREE LANE ENDS (Greater Manchester)
Threekingham (Lincolnshire)
Threlkeld (Cumbria)
THRESHFIELD (North Yorkshire)
Thringarth (Durham)
THRINGSTONE (Leicestershire)
Thrintoft (North Yorkshire)
THROAPHAM (South Yorkshire)
Throckenhalt (Lincolnshire)
THROCKLEY (Tyne & Wear)
Throckmorton (Hereford & Worcester)
THROPHILL (Northumberland)
Thropton (Northumberland)
Througham (Gloucestershire)
Throwley (Kent)
Throwley Forstal (Kent)
Thrumpton (Nottinghamshire)
Thrunscoe (Lincolnshire)
Thrupp (Gloucestershire)
Thrussington (Leicestershire)
Thruxton (Hereford & Worcester)
THRYBERGH (South Yorkshire)
Thulston (Derbyshire)
Thurcaston (Leicestershire)
THURCROFT (South Yorkshire)
Thurgarton (Nottinghamshire)
THURGOLAND (South Yorkshire)
Thurlaston (Leicestershire)
Thurlaston (Warwickshire)
Thurlbear (Somerset)
THURLBY (Lincolnshire)
Thurloxton (Somerset)
THURLSTONE (South Yorkshire)
Thurlwood (Cheshire)
Thurmaston (Leicestershire)
Thurnby (Leicestershire)
Thurnham (Kent)
THURNSCOE (South Yorkshire)
Thursby (Cumbria)
THURSDEN (Lancashire)
Thurstaston (Merseyside)
THURSTON CLOUGH (Greater Manchester)
Thurstonfield (Cumbria)
THURSTONLAND (West Yorkshire)
Thurvaston (Derbyshire)
THWAITE (North Yorkshire)
Thwaite Head (Cumbria)
THWAITES (West Yorkshire)
THWAITES BROW (West Yorkshire)
Thwing (East Riding of Yorkshire)
Tibberton (Gloucestershire)
Tibberton (Hereford & Worcester)
Tibberton (Shropshire)
TIBSHELF (Derbyshire)
Tibthorpe (East Riding of Yorkshire)
Tickenham (Somerset)
TICKHILL (South Yorkshire)
Ticklerton (Shropshire)
TICKNALL (Derbyshire)
Tickton (East Riding of Yorkshire)
Tidbury Green (West Midlands)
Tiddington (Warwickshire)
Tidenham (Gloucestershire)
Tideswell (Derbyshire)
Tidmington (Warwickshire)
TIERS CROSS (Pembrokeshire)
Tile Cross (West Midlands)
TILE HILL (West Midlands)
Tilehouse Green (West Midlands)
Tilham Street (Somerset)
Tillers Green (Gloucestershire)
Tillington (Hereford & Worcester)
Tillington Common (Hereford & Worcester)
TILMANSTONE (Kent)
TILN (Nottinghamshire)
Tilstock (Shropshire)
Tilston (Cheshire)
Tilstone Bank (Cheshire)
Tilstone Fearnall (Cheshire)
Tilton on the Hill (Leicestershire)
Tiltups End (Gloucestershire)
Timberland (Lincolnshire)
TIMBERSBROOK (Cheshire)
Timberscombe (Somerset)
Timble (North Yorkshire)
Timperley (Greater Manchester)
TIMSBURY (Somerset)

See paras. 2 and 4 of the User Guide 2003 if you can't find your place name.

71

TINDALE (Cumbria)
TINDALE CRESCENT (Durham)
TINGLEY (West Yorkshire)
Tinkersley (Derbyshire)
TINSLEY (South Yorkshire)
Tintern Parva (Monmouthshire)
Tintinhull (Somerset)
Tintwistle (Derbyshire)
TIPTON (West Midlands)
TIPTON GREEN (West Midlands)
TIR-Y-FRON (Flintshire)
Tirabad (Powys)
Tirley (Gloucestershire)
TIRPHIL (Caerphilly)
TIRRIL (Cumbria)
Tissington (Derbyshire)
TITHBY (Nottinghamshire)
Titley (Hereford & Worcester)
TITTENSOR (Staffordshire)
Titton (Hereford & Worcester)
Tiverton (Cheshire)
Tivington (Somerset)
TIVY DALE (South Yorkshire)
Tixall (Staffordshire)
TOADHOLE (Derbyshire)
TOADMOOR (Derbyshire)
Tocketts (North Yorkshire)
Tockington (Gloucestershire)
Tockwith (North Yorkshire)
TODBURN (Northumberland)
Toddington (Gloucestershire)
Todenham (Gloucestershire)
Todhills (Cumbria)
TODHILLS (Durham)
Todmorden (West Yorkshire)
TODWICK (South Yorkshire)
Toft (Cheshire)
Toft (Lincolnshire)
Toft (Warwickshire)
TOFT HILL (Durham)
Toft next Newton (Lincolnshire)
TOGSTON (Northumberland)
TOLL BAR (South Yorkshire)
Tolland (Somerset)
TOLLBAR END (West Midlands)
Tollerton (North Yorkshire)
TOLLERTON (Nottinghamshire)
Tomlow (Warwickshire)
TOMPKIN (Staffordshire)
Ton (Monmouthshire)
TON-TEG (Rhondda Cynon Taff)
Tonbridge (Kent)
TONDU (Bridgend)
Tonedale (Somerset)
Tong (Kent)
Tong (Shropshire)
TONG (West Yorkshire)
Tong Green (Kent)
Tong Norton (Shropshire)
TONG STREET (West Yorkshire)
Tonge (Leicestershire)
Tongue End (Lincolnshire)
TONGWYNLAIS (Cardiff)
TONNA (Neath Port Talbot)
TONYPANDY (Rhondda Cynon Taff)
TONYREFAIL (Rhondda Cynon Taff)
TOP OF HEBERS (Greater Manchester)
TOP-Y-RHOS (Flintshire)
Topcliffe (North Yorkshire)
TOPHAM (South Yorkshire)
TOPPINGS (Greater Manchester)
Torksey (Lincolnshire)
Tormarton (Gloucestershire)
TORONTO (Durham)
TORPENHOW (Cumbria)
Torre (Somerset)
Torrisholme (Lancashire)
Tortan (Hereford & Worcester)
Tortworth (Gloucestershire)
Torver (Cumbria)
TORWORTH (Nottinghamshire)
Tosside (North Yorkshire)
Tothill (Lincolnshire)
TOTLEY (South Yorkshire)
TOTLEY BROOK (South Yorkshire)
TOTON (Nottinghamshire)
TOTTERDOWN (Bristol)
TOTTINGTON (Greater Manchester)
TOTTLEWORTH (Lancashire)
Toulston (North Yorkshire)
Toulton (Somerset)

Tovil (Kent)
TOW LAW (Durham)
Tower Hill (Merseyside)
Town End (Cumbria)
Town Green (Lancashire)
Town Head (Cumbria)
Town Head (North Yorkshire)
TOWN KELLOE (Durham)
TOWN LANE (Greater Manchester)
TOWN MOOR (Tyne & Wear)
TOWN OF LOWDON (Merseyside)
Towngate (Cumbria)
Towngate (Lincolnshire)
TOWNHEAD (Cumbria)
TOWNHEAD (South Yorkshire)
Townsend (Somerset)
Townwell (Gloucestershire)
Towthorpe (East Riding of Yorkshire)
Towthorpe (North Yorkshire)
Towton (North Yorkshire)
Toxteth (Merseyside)
Toy's Hill (Kent)
Toynton All Saints (Lincolnshire)
Toynton Fen Side (Lincolnshire)
Toynton St. Peter (Lincolnshire)
Tracebridge (Somerset)
TRAFFORD PARK (Greater Manchester)
Trallong (Powys)
Tranmere (Merseyside)
TRANWELL (Northumberland)
Trap (Carmarthenshire)
Trap's Green (Warwickshire)
TRAWDEN (Lancashire)
Tre Aubrey (Vale of Glamorgan)
Tre-gagle (Monmouthshire)
TRE-GIBBON (Rhondda Cynon Taff)
Tre-Mostyn (Flintshire)
Tre-Vaughan (Carmarthenshire)
Tre-wyn (Monmouthshire)
TREALAW (Rhondda Cynon Taff)
Treales (Lancashire)
Trearddur (Anglesey)
Trearddur Bay (Anglesey)
TREBANOG (Rhondda Cynon Taff)
TREBANOS (Neath Port Talbot)
Treborough (Somerset)
Trecastle (Powys)
Trecwn (Pembrokeshire)
TRECYNON (Rhondda Cynon Taff)
TREDEGAR (Blaenau Gwent)
Tredington (Gloucestershire)
Tredington (Warwickshire)
Tredomen (Powys)
Tredrissi (Pembrokeshire)
Tredunhock (Monmouthshire)
Tredustan (Powys)
TREETON (South Yorkshire)
Trefacca (Powys)
Trefasser (Pembrokeshire)
TREFDRAETH (Anglesey)
Trefeglwys (Powys)
Treffgarne (Pembrokeshire)
Treffgarne Owen (Pembrokeshire)
TREFFOREST (Rhondda Cynon Taff)
Treffynnon (Pembrokeshire)
Trefil (Blaenau Gwent)
TREFLACH WOOD (Shropshire)
Trefnannau (Powys)
Trefnant (Denbighshire)
TREFONEN (Shropshire)
Trefor (Anglesey)
TYNDALL'S PARK (Bristol)
Tregaian (Anglesey)
Tregare (Monmouthshire)
Tregarth (Anglesey)
Tregeiriog (Wrexham)
Tregele (Anglesey)
Treglemais (Pembrokeshire)
Tregoyd (Powys)
Tregynon (Powys)
TREHAFOD (Rhondda Cynon Taff)
TREHARRIS (Merthyr Tydfil)
Treherbert (Carmarthenshire)
TREHERBERT (Rhondda Cynon Taff)
Trelawnyd (Flintshire)
Trelech (Carmarthenshire)
Trelech a'r Betws (Carmarthenshire)
Treleddyd-fawr (Pembrokeshire)
TRELEWIS (Merthyr Tydfil)
Trelleck (Monmouthshire)
Trelleck Grange (Monmouthshire)

Trelogan (Flintshire)
Trelystan (Powys)
Tremeirchion (Denbighshire)
TRENCH (Shropshire)
Trent Port (Lincolnshire)
TRENT VALE (Staffordshire)
TRENTHAM (Staffordshire)
Trentlock (Derbyshire)
Treoes (Vale of Glamorgan)
TREORCHY (Rhondda Cynon Taff)
Trerhyngyll (Vale of Glamorgan)
Trescott (Staffordshire)
Tresham (Gloucestershire)
Treswell (Nottinghamshire)
TRETHOMAS (Caerphilly)
Tretio (Pembrokeshire)
Tretire (Hereford & Worcester)
Tretower (Powys)
TREUDDYN (Flintshire)
TREVALYN (Wrexham)
Trevaughan (Carmarthenshire)
TREVETHIN (Torfaen)
Trevine (Pembrokeshire)
TREVOR (Denbighshire)
Trewalkin (Powys)
Trewent (Pembrokeshire)
Trewern (Powys)
Triangle (West Yorkshire)
TRIERMAIN (Cumbria)
Triffleton (Pembrokeshire)
TRIMDON (Durham)
TRIMDON COLLIERY (Durham)
TRIMDON GRANGE (Durham)
TRIMPLEY (Hereford & Worcester)
TRIMSARAN (Carmarthenshire)
TRINANT (Caerphilly)
Triscombe (Somerset)
TRITLINGTON (Northumberland)
TROEDRHIWFUWCH (Caerphilly)
TROEDYRHIW (Merthyr Tydfil)
Trotshill (Hereford & Worcester)
Trottiscliffe (Kent)
TROUGH GATE (Lancashire)
Troughend (Northumberland)
Troutbeck (Cumbria)
Troutbeck Bridge (Cumbria)
TROWAY (Derbyshire)
TROWELL (Nottinghamshire)
TROY (West Yorkshire)
Trudoxhill (Somerset)
Trull (Somerset)
TRUMFLEET (South Yorkshire)
Trumpet (Hereford & Worcester)
Trunnah (Lancashire)
Trusley (Derbyshire)
Trysull (Staffordshire)
TUCKHILL (Shropshire)
Tudeley (Kent)
TUDHOE (Durham)
Tudorville (Hereford & Worcester)
Tuffley (Gloucestershire)
Tufton (Pembrokeshire)
Tugby (Leicestershire)
Tugford (Shropshire)
TUGHALL (Northumberland)
TUMBLE (Carmarthenshire)
Tumby (Lincolnshire)
Tumby Woodside (Lincolnshire)
Tunbridge Wells (Kent)
Tunstall (East Riding of Yorkshire)
Tunstall (Kent)
TUNSTALL (Lancashire)
Tunstall (North Yorkshire)
TUNSTALL (Staffordshire)
TUNSTALL (Tyne & Wear)
Tunstead (Derbyshire)
TUNSTEAD MILTON (Derbyshire)
Tupsley (Hereford & Worcester)
Tur Langton (Leicestershire)
Turkdean (Gloucestershire)
TURLEYGREEN (Shropshire)
TURN (Lancashire)
Turnastone (Hereford & Worcester)
Turnditch (Derbyshire)
Turner Green (Lancashire)
Turner's Green (Warwickshire)
TURTON BOTTOMS (Greater Manchester)
Tushingham cum Grindley (Cheshire)
Tutbury (Staffordshire)
Tutnall (Hereford & Worcester)
Tutshill (Gloucestershire)

See paras. 2 and 4 of the User Guide 2003 if you can't find your place name.

TUXFORD (Nottinghamshire)
TWEEDMOUTH (Northumberland)
Twemlow Green (Cheshire)
Twenty (Lincolnshire)
TWERTON (Somerset)
Twigworth (Gloucestershire)
Twinhoe (Somerset)
TWISS GREEN (Cheshire)
Twitchen (Shropshire)
TWITHAM (Kent)
Two Dales (Derbyshire)
TWO GATES (Staffordshire)
Twycross (Leicestershire)
Twyford (Leicestershire)
Twyford (Lincolnshire)
Twyford Common (Hereford & Worcester)
TWYN-CARNO (Caerphilly)
Twyn-y-Sheriff (Monmouthshire)
Twyn-yr-Odyn (Vale of Glamorgan)
Twyning (Gloucestershire)
Twyning Green (Gloucestershire)
Twynllanan (Carmarthenshire)
Ty'n-dwr (Denbighshire)
TY'N-Y-BRYN (Rhondda Cynon Taff)
TY'N-Y-COEDCAE (Caerphilly)
TY-CROES (Carmarthenshire)
Tyberton (Hereford & Worcester)
Tyburn (West Midlands)
Tycrwyn (Powys)
Tydd Gote (Lincolnshire)
Tydd St. Mary (Lincolnshire)
TYERSAL (West Yorkshire)
TYLDESLEY (Greater Manchester)
Tyler Hill (Kent)
TYLORSTOWN (Rhondda Cynon Taff)
Tylwch (Powys)
TYN-Y-NANT (Rhondda Cynon Taff)
TYNEMOUTH (Tyne & Wear)
TYNEWYDD (Rhondda Cynon Taff)
Tynygongl (Anglesey)
Tyseley (West Midlands)
Tythegston (Bridgend)
TYTHERINGTON (Cheshire)
TYTHERINGTON (Gloucestershire)
Tytherington (Somerset)

U

Ubley (Somerset)
Uckerby (North Yorkshire)
Uckinghall (Hereford & Worcester)
Uckington (Gloucestershire)
Uckington (Shropshire)
UFFINGTON (Shropshire)
Ufton (Warwickshire)
Ugglebarnby (North Yorkshire)
UGHILL (Derbyshire)
UGTHORPE (North Yorkshire)
Ulcat Row (Cumbria)
Ulceby (Lincolnshire)
Ulceby Cross (Lincolnshire)
Ulceby Skitter (Lincolnshire)
Ulcombe (Kent)
Uldale (Cumbria)
Uley (Gloucestershire)
ULGHAM (Northumberland)
Ullenhall (Warwickshire)
Ullenwood (Gloucestershire)
Ulleskelf (North Yorkshire)
Ullesthorpe (Leicestershire)
ULLEY (South Yorkshire)
Ullingswick (Hereford & Worcester)
ULLOCK (Cumbria)
Ulpha (Cumbria)
Ulrome (East Riding of Yorkshire)
Ulverley Green (West Midlands)
Ulverston (Cumbria)
UMBERLEIGH (Devon)
Under River (Kent)
Underbarrow (Cumbria)
UNDERCLIFFE (West Yorkshire)
UNDERDALE (Shropshire)
UNDERLEY HALL (Cumbria)
Underling Green (Kent)
UNDERWOOD (Nottinghamshire)
Undy (Monmouthshire)
UNSTONE (Derbyshire)
UNSTONE GREEN (Derbyshire)
UNSWORTH (Greater Manchester)
UNTHANK (Cumbria)
UNTHANK (Derbyshire)

UNTHANK (Northumberland)
Unthank End (Cumbria)
UP HOLLAND (Lancashire)
Up Mudford (Somerset)
Upchurch (Kent)
Upcott (Hereford & Worcester)
Upcott (Somerset)
Uphampton (Hereford & Worcester)
Uphill (Somerset)
Upleadon (Gloucestershire)
Upleatham (North Yorkshire)
Uplees (Kent)
Upper Affcot (Shropshire)
UPPER ARLEY (Hereford & Worcester)
UPPER BATLEY (West Yorkshire)
Upper Bentley (Hereford & Worcester)
UPPER BIRCHWOOD (Derbyshire)
UPPER BOAT (Rhondda Cynon Taff)
Upper Brailes (Warwickshire)
Upper Breinton (Hereford & Worcester)
Upper Broadheath (Hereford & Worcester)
UPPER BROUGHTON (Nottinghamshire)
Upper Bush (Kent)
Upper Canada (Somerset)
Upper Catshill (Hereford & Worcester)
Upper Chapel (Powys)
Upper Cheddon (Somerset)
Upper Coberley (Gloucestershire)
Upper Cotton (Staffordshire)
Upper Cound (Shropshire)
UPPER CUDWORTH (South Yorkshire)
UPPER CUMBERWORTH (West Yorkshire)
UPPER CWMTWRCH (Powys)
UPPER DEAL (Kent)
UPPER DENBY (West Yorkshire)
UPPER DENTON (Cumbria)
Upper Dinchope (Shropshire)
Upper Dunsforth (North Yorkshire)
Upper Egleton (Hereford & Worcester)
Upper Elkstone (Staffordshire)
Upper Ellastone (Staffordshire)
Upper End (Derbyshire)
Upper Farmcote (Shropshire)
Upper Framilode (Gloucestershire)
Upper Godney (Somerset)
Upper Green (Monmouthshire)
Upper Grove Common (Hereford & Worcester)
Upper Hackney (Derbyshire)
Upper Halling (Kent)
Upper Harbledown (Kent)
Upper Hardres Court (Kent)
Upper Hardwick (Hereford & Worcester)
UPPER HARTSHAY (Derbyshire)
Upper Hatherley (Gloucestershire)
UPPER HATTON (Staffordshire)
UPPER HAUGH (South Yorkshire)
Upper Hayton (Shropshire)
UPPER HEATON (West Yorkshire)
Upper Helmsley (North Yorkshire)
Upper Hergest (Hereford & Worcester)
UPPER HIENDLEY (West Yorkshire)
Upper Hill (Hereford & Worcester)
Upper Hockenden (Kent)
UPPER HOPTON (West Yorkshire)
Upper Howsell (Hereford & Worcester)
Upper Hulme (Staffordshire)
Upper Kilcott (Gloucestershire)
UPPER KILLAY (Swansea)
UPPER LANDYWOOD (Staffordshire)
Upper Langford (Somerset)
UPPER LANGWITH (Derbyshire)
Upper Leigh (Staffordshire)
Upper Ley (Gloucestershire)
Upper Littleton (Somerset)
UPPER LONGDON (Staffordshire)
Upper Ludstone (Shropshire)
UPPER LYDBROOK (Gloucestershire)
Upper Lyde (Hereford & Worcester)
Upper Lye (Hereford & Worcester)
Upper Maes-coed (Hereford & Worcester)
Upper Midhope (Derbyshire)
Upper Milton (Hereford & Worcester)
Upper Moor (Hereford & Worcester)
UPPER MOOR SIDE (West Yorkshire)
Upper Netchwood (Shropshire)
Upper Nobut (Staffordshire)
Upper Padley (Derbyshire)
Upper Poppleton (North Yorkshire)
UPPER PULLEY (Shropshire)
Upper Quinton (Warwickshire)
Upper Rochford (Hereford & Worcester)

Upper Sapey (Hereford & Worcester)
Upper Shuckburgh (Warwickshire)
Upper Slaughter (Gloucestershire)
UPPER SOUDLEY (Gloucestershire)
Upper Spond (Hereford & Worcester)
Upper Standen (Kent)
Upper Strensham (Hereford & Worcester)
Upper Swell (Gloucestershire)
UPPER TANKERSLEY (South Yorkshire)
UPPER TEAN (Staffordshire)
Upper Threapwood (Cheshire)
Upper Town (Derbyshire)
Upper Town (Durham)
Upper Town (Hereford & Worcester)
Upper Town (Somerset)
Upper Tysoe (Warwickshire)
Upper Upnor (Kent)
UPPER VOBSTER (Somerset)
Upper Weston (Somerset)
Upper Wick (Hereford & Worcester)
Upper Wyche (Hereford & Worcester)
Upperby (Cumbria)
UPPERMILL (Greater Manchester)
Upperthong (West Yorkshire)
UPPERTHORPE (Derbyshire)
Upperthorpe (Lincolnshire)
Uppertown (Derbyshire)
Upperup (Gloucestershire)
Upperwood (Derbyshire)
UPPINGTON (Shropshire)
Upsall (North Yorkshire)
UPSTREET (Kent)
UPTON (Cheshire)
UPTON (Cumbria)
Upton (East Riding of Yorkshire)
Upton (Leicestershire)
Upton (Lincolnshire)
Upton (Merseyside)
Upton (Nottinghamshire)
Upton (Pembrokeshire)
Upton (Somerset)
Upton (Warwickshire)
UPTON (West Yorkshire)
UPTON CHEYNEY (Gloucestershire)
Upton Cressett (Shropshire)
Upton Crews (Hereford & Worcester)
Upton Heath (Cheshire)
UPTON MAGNA (Shropshire)
Upton Noble (Somerset)
Upton Snodsbury (Hereford & Worcester)
Upton St. Leonards (Gloucestershire)
Upton upon Severn (Hereford & Worcester)
Upton Warren (Hereford & Worcester)
Urdimarsh (Hereford & Worcester)
Ure Bank (North Yorkshire)
Urlay Nook (Durham)
Urmston (Greater Manchester)
Urra (North Yorkshire)
USHAW MOOR (Durham)
Usk (Monmouthshire)
Usselby (Lincolnshire)
USWORTH (Tyne & Wear)
Utley (West Yorkshire)
Utterby (Lincolnshire)
Uttoxeter (Staffordshire)
UZMASTON (Pembrokeshire)

V

Valley (Anglesey)
VAN (Caerphilly)
VARTEG (Torfaen)
Vaynor (Merthyr Tydfil)
Velindre (Powys)
Vellow (Somerset)
VENNINGTON (Shropshire)
Vernolds Common (Shropshire)
Vickerstown (Cumbria)
VICTORIA (Blaenau Gwent)
VICTORIA (South Yorkshire)
Vigo (Kent)
VOBSTER (Somerset)
Vowchurch (Hereford & Worcester)
VULCAN VILLAGE (Cheshire)

W

WACKERFIELD (Durham)
Wadborough (Hereford & Worcester)
Waddicar (Merseyside)
Waddingham (Lincolnshire)

See paras. 2 and 4 of the User Guide 2003 if you can't find your place name.

Waddington (Lancashire)
Waddington (Lincolnshire)
Wadeford (Somerset)
WADSHELF (Derbyshire)
WADWORTH (South Yorkshire)
Waen (Denbighshire)
Waen (Powys)
Waen Fach (Powys)
Wagbeach (Shropshire)
WAINFELIN (Torfaen)
Wainfleet All Saints (Lincolnshire)
Wainfleet Bank (Lincolnshire)
Wains Hill (Somerset)
Wainscott (Kent)
WAINSTALLS (West Yorkshire)
WAITBY (Cumbria)
Waithe (Lincolnshire)
Wake Green (West Midlands)
WAKEFIELD (West Yorkshire)
WAL-WEN (Flintshire)
WALBOTTLE (Tyne & Wear)
Walby (Cumbria)
Walcombe (Somerset)
Walcot (Lincolnshire)
Walcot (Shropshire)
Walcot (Warwickshire)
Walcote (Leicestershire)
WALDEN (North Yorkshire)
WALDEN HEAD (North Yorkshire)
WALDEN STUBBS (North Yorkshire)
Walderslade (Kent)
Waldley (Derbyshire)
WALDRIDGE (Durham)
Wales (Somerset)
WALES (South Yorkshire)
Walesby (Lincolnshire)
WALESBY (Nottinghamshire)
Walford (Hereford & Worcester)
Walford (Shropshire)
Walford (Staffordshire)
Walford Heath (Shropshire)
Walgherton (Cheshire)
WALK MILL (Lancashire)
WALKDEN (Greater Manchester)
WALKER (Tyne & Wear)
Walker Fold (Lancashire)
Walker's Green (Hereford & Worcester)
Walker's Heath (West Midlands)
WALKERINGHAM (Nottinghamshire)
Walkerith (Lincolnshire)
Walkington (East Riding of Yorkshire)
WALKLEY (South Yorkshire)
Walkwood (Hereford & Worcester)
WALL (Northumberland)
Wall (Staffordshire)
Wall End (Cumbria)
Wall End (Hereford & Worcester)
WALL HEATH (West Midlands)
Wall Houses (Northumberland)
Wall under Haywood (Shropshire)
Wallasey (Merseyside)
Wallend (Kent)
Waller's Green (Hereford & Worcester)
Wallhead (Cumbria)
WALLINGTON HEATH (West Midlands)
Wallis (Pembrokeshire)
WALLSEND (Tyne & Wear)
Wallthwaite (Cumbria)
WALMER (Kent)
Walmer Bridge (Lancashire)
WALMERSLEY (Greater Manchester)
WALMESTONE (Kent)
Walmley (West Midlands)
Walmley Ash (West Midlands)
Walmsgate (Lincolnshire)
Walpole (Somerset)
Walrow (Somerset)
WALSALL (West Midlands)
WALSALL WOOD (West Midlands)
Walsden (West Yorkshire)
WALSGRAVE On SOWE (West Midlands)
WALSHAW (Greater Manchester)
Walshaw (West Yorkshire)
Walshford (North Yorkshire)
Walters Green (Kent)
Walterston (Vale of Glamorgan)
Walterstone (Hereford & Worcester)
Waltham (Kent)
Waltham (Lincolnshire)
WALTHAM ON THE WOLDS (Leicestershire)
Walton (Cumbria)

WALTON (Derbyshire)
Walton (Leicestershire)
Walton (Merseyside)
Walton (Powys)
Walton (Shropshire)
Walton (Somerset)
Walton (Staffordshire)
Walton (Warwickshire)
WALTON (West Yorkshire)
Walton Cardiff (Gloucestershire)
Walton East (Pembrokeshire)
Walton on the Wolds (Leicestershire)
Walton Park (Somerset)
WALTON WEST (Pembrokeshire)
Walton-in-Gordano (Somerset)
Walton-le-Dale (Lancashire)
Walton-on-the-Hill (Staffordshire)
WALTON-ON-TRENT (Derbyshire)
Walwen (Flintshire)
WALWICK (Northumberland)
Walworth (Durham)
Walworth Gate (Durham)
WALWYN'S CASTLE (Pembrokeshire)
Wambrook (Somerset)
Wampool (Cumbria)
Wanlip (Leicestershire)
Wansford (East Riding of Yorkshire)
Wanshurst Green (Kent)
Wanstrow (Somerset)
Wanswell (Gloucestershire)
Wants Green (Hereford & Worcester)
WAPLEY (Gloucestershire)
Wappenbury (Warwickshire)
Warbreck (Lancashire)
Warburton (Greater Manchester)
Warcop (Cumbria)
Ward End (West Midlands)
Warden (Kent)
WARDEN (Northumberland)
WARDEN LAW (Tyne & Wear)
Wardle (Cheshire)
WARDLE (Greater Manchester)
WARDLEY (Greater Manchester)
Wardlow (Derbyshire)
WARDSEND (Cheshire)
Ware Street (Kent)
Warehorne (Kent)
Waren Mill (Northumberland)
WARENFORD (Northumberland)
WARENTON (Northumberland)
Waresley (Hereford & Worcester)
Wargate (Lincolnshire)
Warham (Hereford & Worcester)
WARK (Northumberland)
WARKWORTH (Northumberland)
Warlaby (North Yorkshire)
Warland (West Yorkshire)
Warleigh (Somerset)
Warley Town (West Yorkshire)
Warmbrook (Derbyshire)
WARMFIELD (West Yorkshire)
Warmingham (Cheshire)
Warmington (Warwickshire)
WARMLEY (Bristol / Gloucestershire)
WARMSWORTH (South Yorkshire)
Warndon (Hereford & Worcester)
Warren (Cheshire)
Warren (Pembrokeshire)
Warren Street (Kent)
WARRINGTON (Cheshire)
Warslow (Staffordshire)
WARSOP (Nottinghamshire)
WARSOP VALE (Nottinghamshire)
Warter (East Riding of Yorkshire)
Warter Priory (East Riding of Yorkshire)
Warthermaske (North Yorkshire)
Warthill (North Yorkshire)
WARTNABY (Leicestershire)
Warton (Lancashire)
Warton (Northumberland)
WARTON (Warwickshire)
Warwick (Cumbria)
Warwick (Warwickshire)
Warwick Bridge (Cumbria)
WARWICKSLAND (Cumbria)
Wasdale Head (Cumbria)
WASH (Derbyshire)
Washbrook (Somerset)
WASHFOLD (North Yorkshire)
Washford (Somerset)
Washingborough (Lincolnshire)

WASHINGTON (Tyne & Wear)
Washwood Heath (West Midlands)
Waskerley (Durham)
Wasperton (Warwickshire)
Wasps Nest (Lincolnshire)
WASS (North Yorkshire)
Watchet (Somerset)
Watchfield (Somerset)
Watchgate (Cumbria)
WATCHILL (Cumbria)
Watendlath (Cumbria)
WATER (Lancashire)
Water Eaton (Staffordshire)
Water End (East Riding of Yorkshire)
WATER FRYSTON (West Yorkshire)
Water Orton (Warwickshire)
WATER STREET (Neath Port Talbot)
Water Yeat (Cumbria)
WATER'S NOOK (Greater Manchester)
Waterend (Cumbria)
Waterfall (Staffordshire)
WATERFOOT (Lancashire)
Waterhead (Cumbria)
WATERHOUSES (Durham)
Waterhouses (Staffordshire)
Wateringbury (Kent)
Waterlane (Gloucestershire)
WATERLOO (Derbyshire)
Waterloo (Hereford & Worcester)
Waterloo (Merseyside)
Waterloo (Pembrokeshire)
Watermillock (Cumbria)
Waterrow (Somerset)
Waters Upton (Shropshire)
WATERSIDE (Cumbria)
WATERSIDE (Lancashire)
WATERSIDE (South Yorkshire)
Waterston (Pembrokeshire)
WATH (North Yorkshire)
WATH UPON DEARNE (South Yorkshire)
WATNALL (Nottinghamshire)
Watton (East Riding of Yorkshire)
WATTSTOWN (Rhondda Cynon Taff)
WATTSVILLE (Caerphilly)
Wauldby (East Riding of Yorkshire)
WAUNARLWYDD (Swansea)
WAUNGRON (Swansea)
WAUNLWYD (Blaenau Gwent)
WAVERBRIDGE (Cumbria)
Waverton (Cheshire)
WAVERTON (Cumbria)
Wavertree (Merseyside)
Wawne (East Riding of Yorkshire)
Waxholme (East Riding of Yorkshire)
Way (Kent)
Way Wick (Somerset)
Wayford (Somerset)
Weacombe (Somerset)
Wear Head (Durham)
Weardley (West Yorkshire)
Weare (Somerset)
Wearne (Somerset)
Weasdale (Cumbria)
WEASTE (Greater Manchester)
Weatheroak Hill (Hereford & Worcester)
Weaverham (Cheshire)
Weaverslake (Staffordshire)
Weaverthorpe (North Yorkshire)
WEBB'S HEATH (Gloucestershire)
Webbington (Somerset)
Webheath (Hereford & Worcester)
Webton (Hereford & Worcester)
Wedding Hall Fold (North Yorkshire)
Weddington (Kent)
WEDDINGTON (Warwickshire)
Wedmore (Somerset)
WEDNESBURY (West Midlands)
WEDNESFIELD (West Midlands)
Weecar (Nottinghamshire)
Weeford (Staffordshire)
Week (Somerset)
Weel (East Riding of Yorkshire)
Weeping Cross (Staffordshire)
Weethley Hamlet (Warwickshire)
Weeton (East Riding of Yorkshire)
Weeton (Lancashire)
Weeton (North Yorkshire)
WEETWOOD (West Yorkshire)
WEIR (Lancashire)
Weirbrook (Shropshire)
WELBECK ABBEY (Nottinghamshire)

See paras. 2 and 4 of the User Guide 2003 if you can't find your place name.

Welbourn (Lincolnshire)
Welburn (North Yorkshire)
Welbury (North Yorkshire)
Welby (Lincolnshire)
Welford-on-Avon (Warwickshire)
Welham (Leicestershire)
WELHAM (Nottinghamshire)
Well (Lincolnshire)
Well (North Yorkshire)
WELL FOLD (West Yorkshire)
Well Hill (Kent)
Welland (Hereford & Worcester)
Welland Stone (Hereford & Worcester)
Wellesbourne (Warwickshire)
Wellesbourne Mountford (Warwickshire)
Wellingore (Lincolnshire)
Wellington (Cumbria)
Wellington (Hereford & Worcester)
WELLINGTON (Shropshire)
Wellington (Somerset)
Wellington Heath (Hereford & Worcester)
Wellington Marsh (Hereford & Worcester)
WELLOW (Nottinghamshire)
Wellow (Somerset)
Wells (Somerset)
Wells Green (Cheshire)
WELLS HEAD (West Yorkshire)
Wellsborough (Leicestershire)
WELSH BICKNOR (Hereford & Worcester)
Welsh End (Shropshire)
Welsh Frankton (Shropshire)
Welsh Hook (Pembrokeshire)
Welsh Newton (Hereford & Worcester)
Welsh St. Donats (Vale of Glamorgan)
Welshampton (Shropshire)
Welshpool (Powys)
WELTON (Cumbria)
Welton (East Riding of Yorkshire)
Welton (Lincolnshire)
Welton le Marsh (Lincolnshire)
Welton le Wold (Lincolnshire)
Welwick (East Riding of Yorkshire)
Wem (Shropshire)
Wembdon (Somerset)
WENNINGTON (Lancashire)
Wensley (Derbyshire)
WENSLEY (North Yorkshire)
WENTBRIDGE (West Yorkshire)
Wentnor (Shropshire)
WENTWORTH (South Yorkshire)
WENTWORTH CASTLE (South Yorkshire)
Wenvoe (Vale of Glamorgan)
Weobley (Hereford & Worcester)
Weobley Marsh (Hereford & Worcester)
Wergs (Staffordshire)
Wern (Powys)
WERN (Shropshire)
Wern-y-gaer (Flintshire)
WERNETH LOW (Greater Manchester)
WERNFFRWD (Swansea)
WERRINGTON (Staffordshire)
Wervin (Cheshire)
Wesham (Lancashire)
WESSINGTON (Derbyshire)
West Aberthaw (Vale of Glamorgan)
WEST ALLERDEAN (Northumberland)
West Appleton (North Yorkshire)
West Ashby (Lincolnshire)
WEST AUCKLAND (Durham)
West Ayton (North Yorkshire)
West Bagborough (Somerset)
WEST BANK (Blaenau Gwent)
West Bank (Cheshire)
West Barkwith (Lincolnshire)
West Barnby (North Yorkshire)
WEST BOLDON (Tyne & Wear)
WEST BOWLING (West Yorkshire)
West Brabourne (Kent)
West Bradford (Lancashire)
West Bradley (Somerset)
WEST BRETTON (West Yorkshire)
WEST BRIDGFORD (Nottinghamshire)
West Briscoe (Durham)
WEST BROMWICH (West Midlands)
West Buckland (Somerset)
WEST BURTON (North Yorkshire)
WEST BUTSFIELD (Durham)
West Butterwick (Lincolnshire)
West Camel (Somerset)
WEST CHEVINGTON (Northumberland)
West Chinnock (Somerset)

West Cliffe (Kent)
West Coker (Somerset)
West Compton (Somerset)
WEST COTTINGWITH (North Yorkshire)
WEST COWICK (East Riding of Yorkshire)
WEST CROSS (Swansea)
West Curthwaite (Cumbria)
West Deeping (Lincolnshire)
West Derby (Merseyside)
WEST DRAYTON (Nottinghamshire)
West Ella (East Riding of Yorkshire)
WEST END (Caerphilly)
West End (Cumbria)
West End (East Riding of Yorkshire)
WEST END (Lancashire)
West End (Lincolnshire)
WEST END (North Yorkshire)
WEST END (Somerset)
WEST END (West Yorkshire)
West Farleigh (Kent)
West Felton (Shropshire)
West Firsby (Lincolnshire)
West Flotmanby (North Yorkshire)
WEST GARFORTH (West Yorkshire)
WEST HADDLESEY (North Yorkshire)
West Hagley (Hereford & Worcester)
WEST HALLAM (Derbyshire)
WEST HALLAM COMMON (Derbyshire)
West Halton (Lincolnshire)
WEST HANDLEY (Derbyshire)
West Harptree (Somerset)
West Hatch (Somerset)
West Heath (West Midlands)
West Heslerton (North Yorkshire)
West Hewish (Somerset)
WEST HOLYWELL (Tyne & Wear)
West Horrington (Somerset)
West Horton (Northumberland)
West Hougham (Kent)
WEST HOUSE (North Yorkshire)
West Howetown (Somerset)
West Huntspill (Somerset)
West Hythe (Kent)
West Keal (Lincolnshire)
West Kingsdown (Kent)
West Kirby (Merseyside)
West Knapton (North Yorkshire)
West Lambrook (Somerset)
West Langdon (Kent)
WEST LAYTON (North Yorkshire)
West Leake (Nottinghamshire)
West Learmouth (Northumberland)
WEST LEES (North Yorkshire)
West Leigh (Somerset)
West Lilling (North Yorkshire)
West Littleton (Gloucestershire)
West Lutton (North Yorkshire)
West Lydford (Somerset)
West Lyng (Somerset)
West Malling (Kent)
West Malvern (Hereford & Worcester)
WEST MARKHAM (Nottinghamshire)
West Marsh (Lincolnshire)
West Marton (North Yorkshire)
WEST MELTON (South Yorkshire)
West Minster (Kent)
West Monkton (Somerset)
WEST MORTON (West Yorkshire)
West Mudford (Somerset)
West Ness (North Yorkshire)
West Newbiggin (Durham)
West Newton (East Riding of Yorkshire)
West Newton (Somerset)
West Peckham (Kent)
WEST PELTON (Durham)
West Pennard (Somerset)
West Quantoxhead (Somerset)
WEST RAINTON (Durham)
West Rasen (Lincolnshire)
West Ravendale (Lincolnshire)
WEST RETFORD (Nottinghamshire)
West Rounton (North Yorkshire)
WEST SCRAFTON (North Yorkshire)
WEST SLEEKBURN (Northumberland)
West Stockwith (Nottinghamshire)
WEST STONESDALE (North Yorkshire)
West Stoughton (Somerset)
WEST STOURMOUTH (Kent)
WEST STREET (Kent)
West Tanfield (North Yorkshire)
West Thorpe (Nottinghamshire)

West Torrington (Lincolnshire)
West Town (Hereford & Worcester)
WEST TOWN (Somerset)
West Weetwood (Northumberland)
West Wick (Somerset)
WEST WILLIAMSTON (Pembrokeshire)
WEST WITTON (North Yorkshire)
WEST WOODBURN (Northumberland)
West Woodlands (Somerset)
West Woodside (Cumbria)
WEST WYLAM (Northumberland)
West Yoke (Kent)
WESTBERE (Kent)
Westborough (Lincolnshire)
Westbrook (Kent)
WESTBURY (Shropshire)
Westbury on Severn (Gloucestershire)
Westbury-on-Trym (Bristol)
Westbury Park (Bristol)
Westbury-sub-Mendip (Somerset)
Westby (Lancashire)
Westcombe (Somerset)
Westcote (Gloucestershire)
Westcott (Somerset)
Wested (Kent)
Westend (Gloucestershire)
WESTEND TOWN (Northumberland)
Westenhanger (Kent)
WESTERDALE (North Yorkshire)
Westerham (Kent)
WESTERHOPE (Tyne & Wear)
WESTERLEIGH (Gloucestershire)
WESTFIELD (Cumbria)
WESTFIELD (Somerset)
Westfield Sole (Kent)
Westfields (Hereford & Worcester)
Westford (Somerset)
Westgate (Durham)
Westgate (Lincolnshire)
WESTGATE HILL (West Yorkshire)
Westgate on Sea (Kent)
Westham (Somerset)
Westhay (Somerset)
WESTHEAD (Lancashire)
Westhide (Hereford & Worcester)
Westholme (Somerset)
Westhope (Hereford & Worcester)
Westhope (Shropshire)
Westhorpe (Lincolnshire)
WESTHOUGHTON (Greater Manchester)
WESTHOUSE (North Yorkshire)
WESTHOUSES (Derbyshire)
WESTLEY (Shropshire)
Westlinton (Cumbria)
Westmarsh (Kent)
WESTNEWTON (Cumbria)
WESTOE (Tyne & Wear)
Weston (Cheshire)
Weston (Hereford & Worcester)
Weston (Lincolnshire)
Weston (North Yorkshire)
WESTON (Nottinghamshire)
WESTON (Shropshire)
Weston (Somerset)
Weston (Staffordshire)
Weston Beggard (Hereford & Worcester)
WESTON COYNEY (Staffordshire)
WESTON HEATH (Shropshire)
Weston Hills (Lincolnshire)
WESTON IN ARDEN (Warwickshire)
Weston Jones (Staffordshire)
Weston Lullingfields (Shropshire)
WESTON RHYN (Shropshire)
Weston Subedge (Gloucestershire)
Weston under Penyard (Hereford & Worcester)
Weston under Wetherley (Warwickshire)
Weston Underwood (Derbyshire)
Weston-in-Gordano (Somerset)
Weston-Super-Mare (Somerset)
Weston-under-Lizard (Staffordshire)
Weston-upon-Trent (Derbyshire)
Westonbirt (Gloucestershire)
Westonzoyland (Somerset)
Westow (North Yorkshire)
Westport (Somerset)
Westra (Vale of Glamorgan)
WESTTHORPE (Derbyshire)
WESTWARD (Cumbria)
Westwell (Kent)
Westwell Leacon (Kent)
Westwick (Durham)

See paras. 2 and 4 of the User Guide 2003 if you can't find your place name.

Westwood (Kent)
WESTWOOD (Nottinghamshire)
WESTWOOD HEATH (West Midlands)
Westwoodside (Lincolnshire)
Wetham Green (Kent)
Wetheral (Cumbria)
Wetherby (West Yorkshire)
WETLEY ROCKS (Staffordshire)
Wettenhall (Cheshire)
Wetton (Staffordshire)
Wetwang (East Riding of Yorkshire)
Wetwood (Staffordshire)
Whaddon (Gloucestershire)
WHALE (Cumbria)
WHALEY (Derbyshire)
WHALEY BRIDGE (Derbyshire)
WHALEY THORNS (Derbyshire)
WHALLEY (Lancashire)
WHALLEY BANKS (Lancashire)
WHALTON (Northumberland)
WHAMLEY (Northumberland)
Whaplode (Lincolnshire)
Whaplode Drove (Lincolnshire)
Wharf (Warwickshire)
Wharfe (North Yorkshire)
Wharles (Lancashire)
WHARNCLIFFE SIDE (South Yorkshire)
Wharram-le-Street (North Yorkshire)
Wharton (Cheshire)
Wharton (Hereford & Worcester)
WHASHTON GREEN (North Yorkshire)
Whasset (Cumbria)
WHASTON (North Yorkshire)
Whatcote (Warwickshire)
WHATELEY (Warwickshire)
WHATLEY (Somerset)
WHATLEY'S END (Gloucestershire)
Whatsole Street (Kent)
WHATSTANDWELL (Derbyshire)
WHATTON (Nottinghamshire)
WHAW (North Yorkshire)
Wheathill (Shropshire)
Wheathill (Somerset)
WHEATLEY (West Yorkshire)
WHEATLEY HILL (Durham)
WHEATLEY HILLS (South Yorkshire)
WHEATLEY LANE (Lancashire)
Wheaton Aston (Staffordshire)
WHEATSHEAF (Wrexham)
Wheddon Cross (Somerset)
Wheelbarrow Town (Kent)
Wheeler's Street (Kent)
Wheelock (Cheshire)
Wheelock Heath (Cheshire)
WHEELTON (Lancashire)
WHELDALE (West Yorkshire)
WHELDRAKE (North Yorkshire)
Whelford (Gloucestershire)
WHELPO (Cumbria)
WHELSTON (Flintshire)
Whenby (North Yorkshire)
Wheston (Derbyshire)
Whetsted (Kent)
Whetstone (Leicestershire)
WHEYRIGG (Cumbria)
Whicham (Cumbria)
Whichford (Warwickshire)
WHICKHAM (Tyne & Wear)
Whin Lane End (Lancashire)
Whinnow (Cumbria)
Whinny Hill (Durham)
Whisby (Lincolnshire)
WHISTON (Merseyside)
WHISTON (South Yorkshire)
WHISTON (Staffordshire)
Whiston Cross (Shropshire)
WHISTON EAVES (Staffordshire)
WHISTON LANE END (Merseyside)
WHITACRE FIELDS (Warwickshire)
Whitbeck (Cumbria)
Whitbourne (Hereford & Worcester)
WHITBURN (Tyne & Wear)
Whitby (Cheshire)
Whitby (North Yorkshire)
Whitbyheath (Cheshire)
Whitchurch (Cardiff)
Whitchurch (Hereford & Worcester)
Whitchurch (Pembrokeshire)
Whitchurch (Shropshire)
Whitchurch (Somerset)
Whitcot (Shropshire)

Whitcott Keysett (Shropshire)
White Ball (Somerset)
White Chapel (Hereford & Worcester)
White Chapel (Lancashire)
WHITE COPPICE (Lancashire)
White End (Hereford & Worcester)
White Kirkley (Durham)
White Ladies Aston (Hereford & Worcester)
White Mill (Carmarthenshire)
WHITE OX MEAD (Somerset)
White Pit (Lincolnshire)
White Stake (Lancashire)
White Stone (Hereford & Worcester)
WHITE-LE-HEAD (Durham)
Whiteacre (Kent)
WHITEACRE HEATH (Warwickshire)
WHITEBIRK (Lancashire)
Whitebrook (Monmouthshire)
Whitechurch (Pembrokeshire)
WHITECLIFFE (Gloucestershire)
WHITECROFT (Gloucestershire)
WHITEFIELD (Greater Manchester)
WHITEFIELD (Somerset)
WHITEFIELD LANE END (Merseyside)
Whitegate (Cheshire)
WHITEHALL (Bristol)
WHITEHAVEN (Cumbria)
Whitehill (Kent)
Whitehouse Common (West Midlands)
Whitelackington (Somerset)
WHITELEY GREEN (Cheshire)
WHITEMOOR (Derbyshire)
WHITEMOOR (Nottinghamshire)
WHITEMOOR (Staffordshire)
Whiteshill (Gloucestershire)
Whitestaunton (Somerset)
Whitewall Corner (North Yorkshire)
WHITEWAY (Gloucestershire)
WHITEWAY (Somerset)
Whitewell (Lancashire)
Whitfield (Gloucestershire)
WHITFIELD (Kent)
Whitfield (Northumberland)
Whitfield Hall (Northumberland)
Whitford (Flintshire)
Whitgreave (Staffordshire)
WHITKIRK (West Yorkshire)
Whitland (Carmarthenshire)
WHITLEY (North Yorkshire)
WHITLEY (South Yorkshire)
WHITLEY BAY (Tyne & Wear)
Whitley Chapel (Northumberland)
Whitley Heath (Staffordshire)
WHITLEY LOWER (West Yorkshire)
Whitley Row (Kent)
Whitlock's End (West Midlands)
WHITLOW (South Yorkshire)
Whitminster (Gloucestershire)
WHITMORE (Staffordshire)
Whitnash (Warwickshire)
Whitney-on-Wye (Hereford & Worcester)
WHITRIGG (Cumbria)
Whitrigglees (Cumbria)
Whitson (Newport)
Whitstable (Kent)
Whittingham (Northumberland)
Whittingslow (Shropshire)
WHITTINGTON (Derbyshire)
Whittington (Gloucestershire)
Whittington (Hereford & Worcester)
WHITTINGTON (Lancashire)
Whittington (Shropshire)
Whittington (Staffordshire)
WHITTINGTON (Warwickshire)
WHITTINGTON MOOR (Derbyshire)
WHITTLE-LE-WOODS (Lancashire)
WHITTLESTONE HEAD (Lancashire)
Whitton (Durham)
Whitton (Lincolnshire)
Whitton (Powys)
WHITTON (Shropshire)
WHITTONSTALL (Northumberland)
WHITWELL (Derbyshire)
Whitwell (North Yorkshire)
Whitwell-on-the-Hill (North Yorkshire)
WHITWICK (Leicestershire)
WHITWOOD (West Yorkshire)
WHITWORTH (Lancashire)
Whixall (Shropshire)
Whixley (North Yorkshire)
Whorlton (Durham)

Whorlton (North Yorkshire)
Whyle (Hereford & Worcester)
Wibdon (Gloucestershire)
WIBSEY (West Yorkshire)
Wibtoft (Warwickshire)
Wichenford (Hereford & Worcester)
Wichling (Kent)
Wick (Gloucestershire)
Wick (Hereford & Worcester)
Wick (Somerset)
Wick (Vale of Glamorgan)
Wick St. Lawrence (Somerset)
Wickenby (Lincolnshire)
WICKERSLEY (South Yorkshire)
WICKHAMBREAUX (Kent)
Wickhamford (Hereford & Worcester)
WICKWAR (Gloucestershire)
WIDDOP (Lancashire)
WIDDRINGTON (Northumberland)
WIDDRINGTON STATION (Tyne & Wear)
WIDE OPEN (Tyne & Wear)
Widmerpool (Nottinghamshire)
WIDNES (Cheshire)
WIGAN (Greater Manchester)
Wigborough (Somerset)
Wiggenstall (Staffordshire)
WIGGINGTON (Shropshire)
Wigginton (North Yorkshire)
WIGGINTON (Staffordshire)
Wigglesworth (North Yorkshire)
Wiggold (Gloucestershire)
Wiggonby (Cumbria)
Wighill (North Yorkshire)
WIGLEY (Derbyshire)
Wigmore (Hereford & Worcester)
Wigmore (Kent)
WIGSLEY (Nottinghamshire)
Wigston (Leicestershire)
Wigston Fields (Leicestershire)
Wigston Parva (Leicestershire)
WIGTHORPE (Nottinghamshire)
Wigtoft (Lincolnshire)
Wigton (Cumbria)
WIGTWIZZLE (South Yorkshire)
Wike (West Yorkshire)
Wilberfoss (East Riding of Yorkshire)
Wilcrick (Newport)
WILDAY GREEN (Derbyshire)
WILDBOARCLOUGH (Cheshire)
Wilden (Hereford & Worcester)
Wildmoor (Hereford & Worcester)
Wildsworth (Lincolnshire)
WILFORD (Nottinghamshire)
Wilkesley (Cheshire)
Wilksby (Lincolnshire)
Willaston (Cheshire)
Willcott (Shropshire)
WILLENHALL (West Midlands)
Willerby (East Riding of Yorkshire)
Willerby (North Yorkshire)
Willersey (Gloucestershire)
Willersley (Hereford & Worcester)
Willesborough (Kent)
Willesborough Lees (Kent)
Willett (Somerset)
WILLEY (Shropshire)
Willey (Warwickshire)
WILLIAMSTOWN (Rhondda Cynon Taff)
Willicote (Warwickshire)
Willingham (Lincolnshire)
Willington (Derbyshire)
WILLINGTON (Durham)
Willington (Kent)
Willington (Warwickshire)
Willington Corner (Cheshire)
WILLINGTON QUAY (Tyne & Wear)
Willitoft (East Riding of Yorkshire)
Williton (Somerset)
Willoughby (Lincolnshire)
Willoughby (Warwickshire)
Willoughby Hills (Lincolnshire)
Willoughby Waterleys (Leicestershire)
Willoughby-on-the-Wolds (Nottinghamshire)
Willoughton (Lincolnshire)
Willow Green (Cheshire)
WILLSBRIDGE (Gloucestershire)
Willtown (Somerset)
Wilmcote (Warwickshire)
Wilmington (Kent)
Wilmington (Somerset)
WILMSLOW (Cheshire)

See paras. 2 and 4 of the User Guide 2003 if you can't find your place name.

WILNECOTE (Staffordshire)
WILPSHIRE (Lancashire)
WILSDEN (West Yorkshire)
Wilsford (Lincolnshire)
WILSHAW (West Yorkshire)
WILSILL (North Yorkshire)
Wilsley Green (Kent)
Wilsley Pound (Kent)
Wilson (Hereford & Worcester)
Wilson (Leicestershire)
Wilsthorpe (Lincolnshire)
Wilton (Cumbria)
Wilton (Hereford & Worcester)
Wilton (North Yorkshire)
WIMBLEBURY (Staffordshire)
Wimpstone (Warwickshire)
Wincanton (Somerset)
Winceby (Lincolnshire)
Wincham (Cheshire)
Winchcombe (Gloucestershire)
Winchet Hill (Kent)
WINCLE (Cheshire)
WINCOBANK (South Yorkshire)
WINDER (Cumbria)
Windermere (Cumbria)
Winderton (Warwickshire)
WINDLEHURST (Greater Manchester)
Windmill (Derbyshire)
Windmill Hill (Somerset)
Windrush (Gloucestershire)
Windsoredge (Gloucestershire)
Windy Arbour (Warwickshire)
WINDY HILL (Wrexham)
Windyharbour (Cheshire)
Winestead (East Riding of Yorkshire)
WINEWALL (Lancashire)
Winford (Somerset)
Winforton (Hereford & Worcester)
WINGATE (Durham)
WINGATES (Greater Manchester)
WINGATES (Northumberland)
WINGERWORTH (Derbyshire)
WINGHAM (Kent)
Wingmore (Kent)
WINKBURN (Nottinghamshire)
Winkhill (Staffordshire)
Winkhurst Green (Kent)
Winksley (North Yorkshire)
WINLATON (Tyne & Wear)
WINLATON MILL (Tyne & Wear)
Winllan (Powys)
Winmarleigh (Lancashire)
Winnall (Hereford & Worcester)
Winnington (Cheshire)
WINSCALES (Cumbria)
Winscombe (Somerset)
Winsford (Cheshire)
Winsford (Somerset)
Winsham (Somerset)
WINSHILL (Staffordshire)
WINSHWEN (Swansea)
Winskill (Cumbria)
Winson (Gloucestershire)
Winson Green (West Midlands)
Winster (Cumbria)
Winster (Derbyshire)
WINSTON (Durham)
Winstone (Gloucestershire)
WINTERBOURNE (Gloucestershire)
Winterburn (North Yorkshire)
Winteringham (Lincolnshire)
Winterley (Cheshire)
WINTERSETT (West Yorkshire)
Winterton (Lincolnshire)
Winthorpe (Lincolnshire)
WINTHORPE (Nottinghamshire)
WINTON (Cumbria)
Winton (North Yorkshire)
Wintringham (North Yorkshire)
WINWICK (Cheshire)
Wirksworth (Derbyshire)
Wirswall (Cheshire)
WISEMAN'S BRIDGE (Pembrokeshire)
WISETON (Nottinghamshire)
Wishanger (Gloucestershire)
WISHAW (Warwickshire)
Wispington (Lincolnshire)
Wissenden (Kent)
Wistanstow (Shropshire)
Wistanswick (Shropshire)
Wistaston (Cheshire)

Wistaston Green (Cheshire)
Wisterfield (Cheshire)
Wiston (Pembrokeshire)
WISTOW (Leicestershire)
WISTOW (North Yorkshire)
Wiswell (Lancashire)
Witcombe (Somerset)
Witham Friary (Somerset)
Witham on the Hill (Lincolnshire)
Withcall (Lincolnshire)
Witherley (Leicestershire)
Withern (Lincolnshire)
Withernsea (East Riding of Yorkshire)
Withernwick (East Riding of Yorkshire)
Witherslack (Cumbria)
Witherslack Hall (Cumbria)
Withiel Florey (Somerset)
Withington (Cheshire)
Withington (Gloucestershire)
Withington (Greater Manchester)
Withington (Hereford & Worcester)
Withington (Shropshire)
Withington (Staffordshire)
Withington Green (Cheshire)
Withington Marsh (Hereford & Worcester)
WITHNELL (Lancashire)
WITHY MILLS (Somerset)
Withybed Green (Hereford & Worcester)
Withybrook (Warwickshire)
Withycombe (Somerset)
WITHYDITCH (Somerset)
Wittersham (Kent)
Witton (West Midlands)
WITTON GILBERT (Durham)
WITTON LE WEAR (Durham)
WITTON PARK (Durham)
Wiveliscombe (Somerset)
Wixford (Warwickshire)
Wixhill (Shropshire)
Wold Newton (East Riding of Yorkshire)
Wold Newton (Lincolnshire)
WOLF HILLS (Northumberland)
Wolf's Castle (Pembrokeshire)
Wolferlow (Hereford & Worcester)
Wolfhampcote (Warwickshire)
Wolfsdale (Pembrokeshire)
WOLLASTON (Shropshire)
WOLLATON (Nottinghamshire)
Wollerton (Shropshire)
WOLLESCOTE (West Midlands)
WOLSELEY (Staffordshire)
WOLSELEY BRIDGE (Staffordshire)
Wolsingham (Durham)
WOLSTANTON (Staffordshire)
WOLSTENHOLME (Greater Manchester)
Wolston (Warwickshire)
Wolsty (Cumbria)
WOLVERHAMPTON (West Midlands)
Wolverley (Hereford & Worcester)
Wolverley (Shropshire)
Wolverton (Kent)
Wolverton (Warwickshire)
Wolvesnewton (Monmouthshire)
Wolvey (Warwickshire)
Wolvey Heath (Warwickshire)
Wolviston (Durham)
Wombleton (North Yorkshire)
WOMBOURNE (Staffordshire)
WOMBWELL (South Yorkshire)
WOMENSWOLD (Kent)
WOMERSLEY (North Yorkshire)
Wonastow (Monmouthshire)
Wood Bevington (Warwickshire)
WOOD EATON (Staffordshire)
WOOD END (Warwickshire)
WOOD END (West Midlands)
Wood Enderby (Lincolnshire)
WOOD HAYES (West Midlands)
Wood Lane (Shropshire)
WOOD LANE (Staffordshire)
WOOD ROW (West Midlands)
Wood Top (Lancashire)
WOODALE (North Yorkshire)
WOODALL (South Yorkshire)
Woodbeck (Nottinghamshire)
WOODBOROUGH (Nottinghamshire)
Woodchester (Gloucestershire)
Woodchurch (Kent)
Woodchurch (Merseyside)
Woodcombe (Somerset)
WOODCOTE (Shropshire)

Woodcote Green (Hereford & Worcester)
Woodcroft (Gloucestershire)
WOODEN (Pembrokeshire)
Woodend (Staffordshire)
Woodford (Gloucestershire)
Woodford (Greater Manchester)
Woodgate (Hereford & Worcester)
WOODGATE (West Midlands)
WOODHALL (North Yorkshire)
WOODHALL HILL (West Yorkshire)
Woodhall Spa (Lincolnshire)
WOODHAM (Durham)
Woodham (Lincolnshire)
Woodhill (Somerset)
WOODHORN (Northumberland)
WOODHORN DEMESNE (Northumberland)
Woodhouse (Leicestershire)
WOODHOUSE (South Yorkshire)
WOODHOUSE (West Yorkshire)
Woodhouse Eaves (Leicestershire)
Woodhouse Green (Staffordshire)
WOODHOUSE MILL (South Yorkshire)
Woodhouses (Cumbria)
WOODHOUSES (Greater Manchester)
Woodhouses (Staffordshire)
WOODKIRK (West Yorkshire)
WOODLAND (Durham)
Woodland (Kent)
Woodland Street (Somerset)
WOODLAND VIEW (South Yorkshire)
Woodlands (Kent)
Woodlands (North Yorkshire)
WOODLANDS (South Yorkshire)
WOODLESFORD (West Yorkshire)
WOODLEY (Greater Manchester)
Woodmancote (Gloucestershire)
Woodmancote (Hereford & Worcester)
Woodmansey (East Riding of Yorkshire)
Woodmill (Staffordshire)
WOODNESBOROUGH (Kent)
WOODNOOK (Nottinghamshire)
Woodplumpton (Lancashire)
Woodrow (Hereford & Worcester)
Woodseaves (Shropshire)
Woodseaves (Staffordshire)
WOODSETTS (South Yorkshire)
WOODSIDE (Cumbria)
Woodside Green (Kent)
WOODTHORPE (Derbyshire)
Woodthorpe (Leicestershire)
Woodthorpe (Lincolnshire)
Woodvale (Merseyside)
WOODVILLE (Derbyshire)
WOODTOWN (Devon)
Woodwall Green (Staffordshire)
Woofferton (Shropshire)
Wookey (Somerset)
Wookey Hole (Somerset)
WOOLAGE GREEN (Kent)
WOOLAGE VILLAGE (Kent)
WOOLASTON (Gloucestershire)
Woolaston Common (Gloucestershire)
Woolavington (Somerset)
Woolcotts (Somerset)
WOOLDALE (West Yorkshire)
Wooler (Northumberland)
WOOLEY BRIDGE (Derbyshire)
WOOLFOLD (Greater Manchester)
Woolhope (Hereford & Worcester)
WOOLLARD (Somerset)
WOOLLEY (Derbyshire)
Woolley (Somerset)
WOOLLEY (West Yorkshire)
Woolmere Green (Hereford & Worcester)
Woolmerston (Somerset)
Woolminstone (Somerset)
Woolpack (Kent)
Woolscott (Warwickshire)
WOOLSINGTON (Tyne & Wear)
WOOLSTASTON (Shropshire)
Woolsthorpe (Lincolnshire)
WOOLSTON (Cheshire)
Woolston (Shropshire)
Woolston (Somerset)
Woolstone (Gloucestershire)
Woolton (Merseyside)
Woolverton (Somerset)
Woonton (Hereford & Worcester)
Wooperton (Northumberland)
Woore (Shropshire)
Wootton (Hereford & Worcester)

See paras. 2 and 4 of the User Guide 2003 if you can't find your place name.

77

Wootton (Kent)
Wootton (Lincolnshire)
Wootton (Shropshire)
Wootton (Staffordshire)
Wootton Courtenay (Somerset)
Wootton Wawen (Warwickshire)
Worcester (Hereford & Worcester)
WORDSLEY (West Midlands)
Worfield (Shropshire)
WORKINGTON (Cumbria)
WORKSOP (Nottinghamshire)
Worlaby (Lincolnshire)
Worle (Somerset)
Worleston (Cheshire)
Wormald Green (North Yorkshire)
Wormbridge (Hereford & Worcester)
Wormelow Tump (Hereford & Worcester)
Wormhill (Derbyshire)
Wormhill (Hereford & Worcester)
Wormington (Gloucestershire)
Worminster (Somerset)
Wormleighton (Warwickshire)
WORMLEY HILL (South Yorkshire)
Wormshill (Kent)
Wormsley (Hereford & Worcester)
WORRALL (South Yorkshire)
WORRALL HILL (Gloucestershire)
WORSBROUGH (South Yorkshire)
WORSBROUGH BRIDGE (South Yorkshire)
WORSBROUGH DALE (South Yorkshire)
WORSLEY (Greater Manchester)
WORSLEY MESNES (Greater Manchester)
WORSTHORNE (Lancashire)
Worston (Lancashire)
WORTH (Kent)
Worth (Somerset)
Worthen (Shropshire)
WORTHENBURY (Wrexham)
WORTHINGTON (Leicestershire)
Worthybrook (Monmouthshire)
WORTLEY (South Yorkshire)
WORTLEY (West Yorkshire)
WORTON (North Yorkshire)
Wotherton (Shropshire)
Wotton-under-Edge (Gloucestershire)
Wouldham (Kent)
Woundale (Shropshire)
Wragby (Lincolnshire)
WRAGBY (West Yorkshire)
WRANGBROOK (West Yorkshire)
Wrangle (Lincolnshire)
Wrangle Common (Lincolnshire)
Wrangle Lowgate (Lincolnshire)
Wrangway (Somerset)
Wrantage (Somerset)
Wrawby (Lincolnshire)
WRAXALL (Somerset)
WRAY (Lancashire)
Wray Castle (Cumbria)
WRAYTON (Lancashire)
Wrea Green (Lancashire)
Wreaks End (Cumbria)
Wreay (Cumbria)
WREKENTON (Tyne & Wear)
Wrelton (North Yorkshire)
Wrenbury (Cheshire)
Wrench Green (North Yorkshire)
WRENTHORPE (West Yorkshire)
WRENTNALL (Shropshire)

WRESSLE (East Riding of Yorkshire)
Wressle (Lincolnshire)
WREXHAM (Wrexham)
WRIBBENHALL (Hereford & Worcester)
Wrickton (Shropshire)
WRIGHTINGTON BAR (Lancashire)
Wrinehill (Staffordshire)
Wrington (Somerset)
WRITHLINGTON (Somerset)
Wrockwardine (Shropshire)
Wroot (Lincolnshire)
WROSE (West Yorkshire)
Wrotham (Kent)
Wrotham Heath (Kent)
Wrottesley (Staffordshire)
Wroxall (Warwickshire)
WROXETER (Shropshire)
Wyaston (Derbyshire)
Wyberton (Lincolnshire)
Wybunbury (Cheshire)
Wychbold (Hereford & Worcester)
Wychnor (Staffordshire)
Wyck Rissington (Gloucestershire)
Wycliffe (Durham)
WYCOLLER (Lancashire)
WYCOMB (Leicestershire)
Wye (Kent)
Wyesham (Monmouthshire)
Wyfordby (Leicestershire)
WYKE (Shropshire)
WYKE (West Yorkshire)
Wyke Champflower (Somerset)
Wykeham (North Yorkshire)
Wyken (Shropshire)
WYKEN (West Midlands)
Wykey (Shropshire)
Wykin (Leicestershire)
WYLAM (Northumberland)
Wylde Green (West Midlands)
Wymeswold (Leicestershire)
Wymondham (Leicestershire)
WYNDHAM (Bridgend)
Wynds Point (Hereford & Worcester)
Wyre Piddle (Hereford & Worcester)
Wysall (Nottinghamshire)
Wyson (Hereford & Worcester)
Wythall (Hereford & Worcester)
Wythburn (Cumbria)
Wythenshawe (Greater Manchester)
Wythop Mill (Cumbria)
Wyton (East Riding of Yorkshire)
Wyville (Lincolnshire)

Y

Y Gyffylliog (Denbighshire)
Y NANT (Wrexham)
Yaddlethorpe (Lincolnshire)
Yafforth (North Yorkshire)
Yalding (Kent)
YANWATH (Cumbria)
Yanworth (Gloucestershire)
Yapham (East Riding of Yorkshire)
Yarborough (Somerset)
Yarburgh (Lincolnshire)
Yardley (West Midlands)
Yardley Wood (West Midlands)
Yardro (Powys)
Yarford (Somerset)

Yarkhill (Hereford & Worcester)
Yarley (Somerset)
Yarlington (Somerset)
YARLSBER (North Yorkshire)
Yarm (North Yorkshire)
Yarnfield (Staffordshire)
Yarpole (Hereford & Worcester)
Yarrow (Somerset)
Yarsop (Hereford & Worcester)
YATE (Gloucestershire)
Yatton (Hereford & Worcester)
Yatton (Somerset)
Yawthorpe (Lincolnshire)
Yazor (Hereford & Worcester)
YEADON (West Yorkshire)
Yealand Conyers (Lancashire)
Yealand Redmayne (Lancashire)
Yealand Storrs (Lancashire)
Yearby (North Yorkshire)
YEARNGILL (Cumbria)
Yearsley (North Yorkshire)
Yeaton (Shropshire)
Yeaveley (Derbyshire)
Yeavering (Northumberland)
Yedingham (North Yorkshire)
Yenston (Somerset)
Yeo Mill (Somerset)
Yeovil (Somerset)
Yeovil Marsh (Somerset)
Yeovilton (Somerset)
YERBESTON (Pembrokeshire)
Yetlington (Northumberland)
Yew Green (Warwickshire)
YEWS GREEN (West Yorkshire)
Yieldingtree (Hereford & Worcester)
YNYSBOETH (Rhondda Cynon Taff)
YNYSDDU (Caerphilly)
YNYSFORGAN (Swansea)
YNYSHIR (Rhondda Cynon Taff)
YNYSMAERDY (Rhondda Cynon Taff)
YNYSMEUDWY (Neath Port Talbot)
YNYSTAWE (Swansea)
YNYSWEN (Powys)
YNYSWEN (Rhondda Cynon Taff)
YNYSYBWL (Rhondda Cynon Taff)
YOCKENTHWAITE (North Yorkshire)
YOCKLETON (Shropshire)
Yokefleet (East Riding of Yorkshire)
YORK (Lancashire)
York (North Yorkshire)
Yorkletts (Kent)
YORKLEY (Gloucestershire)
Yorton Heath (Shropshire)
Youlgreave (Derbyshire)
Youlthorpe (East Riding of Yorkshire)
Youlton (North Yorkshire)
Yoxall (Staffordshire)
Ysceifiog (Flintshire)
YSTALYFERA (Powys)
YSTRAD (Rhondda Cynon Taff)
YSTRAD MYNACH (Caerphilly)
Ystrad-ffyn (Carmarthenshire)
YSTRADFELLTE (Powys)
YSTRADGYNLAIS (Powys)
Ystradowen (Vale of Glamorgan)

Z

Zouch (Nottinghamshire)

See paras. 2 and 4 of the User Guide 2003 if you can't find your place name.

CON 29M (2003) COAL MINING SEARCH

A cheque with the appropriate fee MUST be sent with the printed form. A location plan should be included with all search requests to avoid delays. Users should retain a copy of the form and the location plan.

The Coal Authority Mining Reports 200 Lichfield Lane Mansfield Nottinghamshire NG18 4RG **DX 716176 MANSFIELD 5**	(For Coal Authority Use Only)

Complete each applicable box below. Use type or block LETTERS.

	Notes
Reply To: DX	Enter the name, firm/organisation and DX number to which the report is to be returned. If you are not a member of the DX system please enter your name, firm/organisation and full address including postcode to which the report is to be returned. Please use DX in preference to post where possible.
Email (if applicable):	Only insert your email when you wish the report to be returned by email in secure .pdf format.
Tel No:	Insert your telephone number so that the Authority may contact you regarding this search.
Account No. (if applicable):	See paragraph 16 of the User Guide 2003.
Report Type Requested (tick box): Residential Property Search ☐ Non-Residential Property or Development Site Search ☐	Ensure that the correct fee is enclosed for the report type requested (see Notes overleaf). It is inappropriate to request a residential property search for non-residential property or development sites, and any such requests will be returned since the user will not be satisfying due diligence requirements. A Residential Property Search includes enquiries 1–9 overleaf. A Non-Residential or Development Site Search includes enquiries 1–12 overleaf.
Expedited Search Required (delete as applicable): **YES/NO** Fax No. (only for Expedited Searches):	Expedited Search Reports are returned by Fax within 48 hours. Choosing "YES" constitutes an undertaking to pay the additional fee to the Authority within seven days (see Notes overleaf).
Request Date:	Insert the date on which this search form is sent.
Address of the Land/Property: NLPG UPRN Address 1 Address 2 Street Locality Town/Village County Postcode	Enter the National Land and Property Gazetteer Unique Property Reference Number where this is known, otherwise leave blank. The Address 1 and Address 2 fields may be used for house and flat names and numbers. This address format complies with BS 7666. Include the complete address if possible. Always include the postcode where the property has a postal address. Where the land or property does not have a postal address, use any of the fields provided to include a description sufficient for the property to be identified.
Location Plan Enclosed (delete as applicable): **YES/NO**	Searches may be returned unanswered or delayed if a plan is not enclosed (see Notes overleaf).
Customer Ref:	Insert your own file reference number.

TERMS AND CONDITIONS
These enquiries are made and the replies prepared in accordance with the Coal Authority's Terms & Conditions 2003, User Guide 2003 and Law Society's Guidance Notes 2003.

ENQUIRIES AND NOTES
The Residential and Non-Residential or Development Site Search Enquiries and some notes regarding the use of this search are reproduced overleaf.

CON 29M (2003) COAL MINING SEARCH

RESIDENTIAL PROPERTY SEARCH (enquiries 1–9)

1. Past underground coal mining

Is the property within the zone of likely physical influence on the surface of past underground coal workings? If yes, indicate the seams involved and approximate date of working.

2. Present underground coal mining

Is the property within the zone of likely physical influence on the surface of present underground coal workings? If yes, indicate the seams involved.

3. Future underground coal mining

(a) Is the property within any area for which an application for a licence to mine coal by underground methods is awaiting determination by the Coal Authority?

(b) Is the property within any area for which a licence to mine coal by underground methods is extant? If yes, when was the licence granted?

(c) Is the property within the zone of likely physical influence on the surface of current planned future underground coal workings? If yes, indicate the seams involved and approximate date of working.

(d) Has any notice of proposals relating to underground coal mining operations been given under s.46 of the Coal Mining Subsidence Act 1991? If yes, supply the date and details of the last such notice.

4. Shafts and adits

Are there any shafts and adits or other entries to underground coal mine workings within the property or within 20 metres of the boundary of the property? If yes, supply a plan showing the approximate recorded location and any relevant information regarding such shafts, adits or entries including, where available, details of any treatment carried out.

5. Surface geology

Is there any fault or other line of weakness at the surface which is known to the Coal Authority to affect the stability of the property?

6. Past opencast coal mining

Is the property situated within an opencast site boundary from which coal has been extracted in the past by opencast methods?

7. Present opencast coal mining

Is the property within 200 metres of an opencast site boundary within which coal is being extracted by opencast methods?

8. Future opencast coal mining

(a) Is the property within 800 metres of any area for which an application for a licence to extract coal by opencast methods is awaiting determination by the Coal Authority?

(b) Is the property within 800 metres of any area for which a licence to extract coal by opencast methods is extant? If yes, when was the licence granted?

9. Subsidence

(a) Has any damage notice or claim for alleged coal mining subsidence damage to the property been given, made or pursued since 1 January 1984? If yes, supply the date of such notice or claim.

(b) In respect of any such notice or claim has the responsible person given notice agreeing that there is a remedial obligation or otherwise accepted that a claim would lie against him?

(c) In respect of any such notice or acceptance has the remedial obligation or claim been discharged? If yes, state whether such remedial obligation or claim was discharged by repair or payment, or a combination thereof.

(d) Does any current 'Stop Notice' concerning the deferment of remedial works or repairs affect the property? If yes, supply the date of the notice.

(e) Has any request been made to execute preventive works under s.33 of the 1991 Act which would prevent the occurrence or reduce the extent of subsidence damage to any buildings, structures or works and, if yes, has any person withheld consent or failed to comply with any such request to execute preventive works?

NON-RESIDENTIAL PROPERTY OR DEVELOPMENT SITE SEARCH (including enquiries 1–9 above and 10–12 below)

10. Withdrawal of support

(a) Does the land lie within an area in respect of which a notice of entitlement to withdraw support has been published? If yes, supply the date of the notice.

(b) Does the land lie within an area in respect of which a revocation notice has been given under s.41 of the Coal Industry Act 1994? If yes, supply the date of the notice.

11. Working facilities orders

Is the property affected by an order in respect of the working of coal under the Mines (Working Facilities and Support) Acts 1923 and 1966 or any statutory modification or amendment thereof? If yes, supply the date and title of the order.

12. Payments to owners of former copyhold land

(a) Has any relevant notice which may affect the property been given?

(b) If yes, has any notice of retained interests in coal and coal mines been given?

(c) If yes, has any acceptance notice or rejection notice been served?

(d) If any such acceptance notice has been served, has any compensation been paid to a claimant?

NOTES

(A) Advice and information on coal mining searches

For information on procedure, turnaround and current fees, or further to a specific search, contact the Coal Authority Mining Reports Helpline on 0845 762 6848 or by email on miningreports@coal.gov.uk or visit www.coalminingreports.co.uk. There is a higher fee for the Non-Residential or Development Site Search and an additional fee for Expedited Searches.

(B) Refunds

No refund or transfer of any fee (or part thereof) will be made once a search has been registered and a location of the property attempted.

(C) Expedited Searches

Select 'YES' against Expedited Search and add your fax number in full. Fax the form to the Authority on 01623 638338. Doing so constitutes an undertaking to send the form with a plan and the additional fee through the post or DX to the Authority within seven days. Searches having incorrect fees or search details will be returned uncompleted.

(D) Affected areas

An electronic Directory of Places with a 'search by postcode' facility can be found on the Authority's website at www.coalmining reports.co.uk.

(E) Ordering coal mining reports

Reports can be ordered and returned by post, telephone, fax, email, and Internet (through www.coalminingreports.co.uk and NLIS channels).

(F) Terms and Conditions, User Guide and Law Society Guidance

These Notes should be read in conjunction with the Coal Authority's Terms and Conditions 2003, User Guide 2003, and Law Society's Guidance Notes 2003. These have been approved by the Law Society and apply to all methods of requesting and receiving a Coal Mining Report from the Authority. Copies of these documents can be found on the Coal Authority's website at www.coalminingreports.co.uk.

(G) Plans

If a plan is not submitted there may be difficulties and delays in identifying the property or its extent, in which case the Coal Authority may request that a plan be supplied before the search can be completed.

The Law Society